Curiosities of the Sky

Curiosities of the Sky

Garrett P. Serviss

ÆGYPAN PRESS

Curiosities of the Sky was first published in 1909.

Curiosities of the Sky
A publication of
ÆGYPAN PRESS

www.aegypan.com

Preface

What Froude says of history is true also of astronomy: it is the most impressive where it transcends explanation. It is not the mathematics of astronomy, but the wonder and the mystery that seize upon the imagination. The calculation of an eclipse owes all its prestige to the sublimity of its data; the operation, in itself, requires no more mental effort than the preparation of a railway time-table.

The dominion which astronomy has always held over the minds of men is akin to that of poetry; when the former becomes merely instructive and the latter purely didactic, both lose their power over the imagination. Astronomy is known as the oldest of the sciences, and it will be the longest-lived because it will always have arcana that have not been penetrated.

Some of the things described in this book are little known to the average reader, while others are well known; but all possess the fascination of whatever is strange, marvelous, obscure, or mysterious — magnified, in this case, by the portentous scale of the phenomena.

The idea of the author is to tell about these things in plain language, but with as much scientific accuracy as plain language will permit, showing the wonder that is in them without getting away from the facts. Most of them have hitherto been discussed only in technical form, and in treatises that the general public seldom sees and never reads.

Among the topics touched upon are:

- The strange unfixedness of the "fixed stars," the vast migrations of the suns and worlds constituting the universe.
- The slow passing out of existence of those collocations of stars which for thousands of years have formed famous "constellations," preserving the memory of mythological heroes and heroines, and perhaps of otherwise unrecorded history.
- The tendency of stars to assemble in immense clouds, swarms, and clusters.
- The existence in some of the richest regions of the universe of absolutely black, starless gaps, deeps, or holes, as if one were looking out of a window into the murkiest night.
- The marvelous phenomena of new, or temporary, stars, which appear as suddenly as conflagrations, and often turn into something else as eccentric as themselves.
- The amazing forms of the "whirlpool," "spiral," "pinwheel," and "lace," or "tress," nebulæ.
- The strange surroundings of the sun, only seen in particular circumstances, but evidently playing a constant part in the daily phenomena of the solar system.
- The mystery of the Zodiacal Light and the Gegenschein.
- The extraordinary transformations undergone by comets and their tails.
- The prodigies of meteorites and masses of stone and metal fallen from the sky.
- The cataclysms that have wrecked the moon.
- The problem of life and intelligence on the planet Mars.
- The problematical origin and fate of the asteroids.
- The strange phenomena of the auroral lights.

An attempt has been made to develop these topics in an orderly way, showing their connection, so that the reader may obtain a broad general view of the chief mysteries and problems of astronomy, and an idea of the immense field of discovery which still lies, almost unexplored, before it.

The Windows of Absolute Night

*T*o most minds mystery is more fascinating than science. But when science itself leads straight up to the borders of mystery and there comes to a dead stop, saying, "At present I can no longer see my way," the force of the charm is redoubled. On the other hand, the illimitable is no less potent in mystery than the invisible, whence the dramatic effect of Keats' "stout Cortez" staring at the boundless Pacific while all his men look at each other with a wild surmise, "silent upon a peak in Darien." It is with similar feelings that the astronomer regards certain places where from the peaks of the universe his vision seems to range out into endless empty space. He sees there the shore of his little isthmus, and, beyond, unexplored immensity.

The name, "coal-sacks," given to these strange voids is hardly descriptive. Rather they produce upon the mind the effect of blank windows in a lonely house on a pitch-dark night, which, when looked at from the brilliant interior, become appalling in their rayless murk. Infinity seems to acquire a new meaning in the presence of these black openings in the sky, for as one continues to gaze it loses its purely metaphysical quality and becomes a kind of entity, like the ocean. The observer is conscious that he can actually *see* the beginning of its ebon depths, in which the visible universe appears to float like an enchanted island, resplendent within with lights and life and gorgeous spectacles, and encircled with screens of crowded stars, but with its dazzling vistas

ending at the fathomless sea of pure darkness which encloses
all.

The Galaxy, or Milky Way, surrounds the borders of our
island in space like a stellar garland, and when openings
appear in it they are, by contrast, far more impressive than
the general darkness of the interstellar expanse seen in other
directions. Yet even that expanse is not everywhere equally
dark, for it contains gloomy deeps discernable with careful
watching. Here, too, contrast plays an important part, though
less striking than within the galactic region. Some of Sir
William Herschel's observations appear to indicate an association between these tenebrious spots and neighboring star
clouds and nebulæ. It is an illuminating bit of astronomical
history that when he was sweeping the then virgin heavens
with his great telescopes he was accustomed to say to his sister
who, notebook in hand, waited at his side to take down his
words, fresh with the inspiration of discovery: "Prepare to
write; the nebulæ are coming; here space is vacant."

The most famous of the "coal-sacks," and the first to be
brought to general attention before astronomers had awakened to the significance of such things, lies adjacent to the
"Southern Cross," and is truly an amazing phenomenon. It
is not alone the conspicuousness of this celestial vacancy,
opening suddenly in the midst of one of the richest parts of
the Galaxy, that has given it its fame, but quite as much the
superstitious awe with which it was regarded by the early
explorers of the South Seas. To them, as well as to those who
listened in rapt wonder to their tales, the "Coal-sack" seemed
to possess some occult connection with the mystic "Cross."
In the eyes of the sailors it was not a vacancy so much as a
sable reality in the sky, and as, shuddering, they stared at it,
they piously crossed themselves. It was another of the magical
wonders of the unknown South, and as such it formed the
basis of many a "wild surmise" and many a sea-dog's yarn.
Scientific investigation has not diminished its prestige, and
today no traveler in the southern hemisphere is indifferent
to its fascinating strangeness, while some find it the most
impressive spectacle of the Antarctic heavens.

All around, up to the very edge of the yawning gap, the
sheen of the Milky Way is surpassingly glorious; but there, as

if in obedience to an almighty edict, everything vanishes. A single faint star is visible within the opening, producing a curious effect upon the sensitive spectator, like the sight of a tiny islet in the midst of a black, motionless, waveless tarn. The dimensions of the lagoon of darkness, which is oval or pear-shaped, are eight degrees by five, so that it occupies a space in the sky about one hundred and thirty times greater than the area of the full moon. It attracts attention as soon as the eye is directed toward the quarter where it exists, and by virtue of the rarity of such phenomena it appears a far greater wonder than the drifts of stars that are heaped around it. Now that observatories are multiplying in the southern hemisphere, the great austral "Coal-sack" will, no doubt, receive attention proportioned to its importance as one of the most significant features of the sky. Already at the Sydney Observatory photographs have shown that the southern portion of this Dead Sea of Space is not quite "bottomless," although its northern part defies the longest sounding lines of the astronomer.

There is a similar, but less perfect, "coal-sack" in the northern hemisphere, in the constellation of "The Swan," which, strange to say, also contains a well-marked figure of a cross outlined by stars. This gap lies near the top of the cross-shaped figure. It is best seen by averted vision, which brings out the contrast with the Milky Way, which is quite brilliant around it. It does not, however, exercise the same weird attraction upon the eye as the southern "Coal-sack," for instead of looking like an absolute void in the sky, it rather appears as if a canopy of dark gauze had been drawn over the stars. We shall see the possible significance of this appearance later.

Just above the southern horizon of our northern middle latitudes, in summer, where the Milky Way breaks up into vast sheets of nebulous luminosity, lying over and between the constellations Scorpio and Sagittarius, there is a remarkable assemblage of "coal-sacks," though none is of great size. One of them, near a conspicuous star-cluster in Scorpio, M80, is interesting for having been the first of these strange objects noted by Herschel. Probably it was its nearness to M80 which suggested to his mind the apparent connection of such vacancies with star-clusters which we have already mentioned.

But the most marvelous of the "coal-sacks" are those that have been found by photography in Sagittarius. One of Barnard's earliest and most excellent photographs includes two of them, both in the star-cluster M8. The larger, which is roughly rectangular in outline, contains one little star, and its smaller neighbor is lune-shaped — surely a most singular form for such an object. Both are associated with curious dark lanes running through the clustered stars like trails in the woods. Along the borders of these lanes the stars are ranked in parallel rows, and what may be called the bottoms of the lanes are not entirely dark, but pebbled with faint stellar points. One of them which skirts the two dark gaps and traverses the cluster along its greatest diameter is edged with lines of stars, recalling the alignment of the trees bordering a French highway. This *road of stars* cannot be less than many billions of miles in length!

All about the cluster the bed of the Galaxy is strangely disturbed, and in places nearly denuded, as if its contents had been raked away to form the immense stack and the smaller accumulations of stars around it. The well-known "Trifid Nebula" is also included in the field of the photograph, which covers a truly marvelous region, so intricate in its mingling of nebulæ, star-clusters, star-swarms, star-streams, and dark vacancies that no description can do it justice. Yet, chaotic as it appears, there is an unmistakable suggestion of unity about it, impressing the beholder with the idea that all the different parts are in some way connected, and have not been fortuitously thrown together. Miss Agnes M. Clerke made the striking remark that the dusky lanes in M8 are exemplified on the largest scale in the great rift dividing the Milky Way, from Cygnus in the northern hemisphere all the way to the "Cross" in the southern. Similar lanes are found in many other clusters, and they are generally associated with flanking rows of stars, resembling in their arrangement the thickset houses and villas along the roadways that traverse the approaches to a great city.

But to return to the black gaps. Are they really windows in the star-walls of the universe? Some of them look rather as if they had been made by a shell fired through a luminous target, allowing the eye to range through the hole into the void space beyond. If science is discretely silent about these things, what can the more venturesome and less responsible imagination suggest? Would a huge "runaway sun," like Arcturus, for instance, make such an opening if it should pass like a projectile through the Milky Way? It is at least a stimulating inquiry. Being probably many thousands of times more massive than the galactic stars, such a stellar missile would not be stopped by them, though its direction of flight might be altered. It would drag the small stars lying close to its course out of their spheres, but the ultimate tendency of its attraction would be to sweep them round in its wake, thus producing rather a star-swarm than a vacancy. Those that were very close to it might be swept away in its rush and become its satellites, careering away with it in its flight into outer space; but those that were farther off, and they would, of course, greatly outnumber the nearer ones, would tend inward from all sides toward the line of flight, as dust and leaves collect behind a speeding motor (though the forces operating would be different), and would fill up the hole, if hole it were. A swarm thus collected should be rounded in outline and bordered with a relatively barren ring from which the stars had been "sucked" away. In a general sense the M8 cluster answers to this description, but even if we undertook to account for its existence by a supposition like the above, the black gaps would remain unexplained, unless one could make a further draft on the imagination and suggest that the stars had been thrown into a vast eddy, or system of eddies, whose vortices appear as dark holes. Only a maelstromlike motion could keep such a funnel open, for without regard to the impulse derived from the projectile, the proper motions of the stars themselves would tend to fill it. Perhaps some other cause of the whirling motion may be found. As we shall see when we come to the spiral nebulæ, gyratory movements are exceedingly prevalent throughout the universe, and the structure of the Milky Way is everywhere suggestive of them. But this is hazardous sport even for the imagination — to play with *suns* as if they were but thistle-down in the wind or corks in a mill-race.

Another question arises: What is the thickness of the hedge of stars through which the holes penetrate? Is the depth of the openings proportionate to their width? In other words, is the Milky Way round in section like a rope, or flat and thin like a ribbon? The answer is not obvious, for we have little or no information concerning the relative distances of the faint galactic stars. It would be easier, certainly, to conceive of openings in a thin belt than in a massive ring, for in the first case they would resemble mere rifts and breaks, while in the second they would be like wells or bore-holes. Then, too, the fact that the Milky Way is not a *continuous* body but is made up of stars whose actual distances apart is great, offers another quandary; persistent and sharply bordered apertures in such an assemblage are *a priori* as improbable, if not impossible, as straight, narrow holes running through a swarm of bees.

The difficulty of these questions indicates one of the reasons why it has been suggested that the seeming gaps, or many of them, are not openings at all, but opaque screens cutting off the light from stars behind them. That this is quite possible in some cases is shown by Barnard's later photographs, particularly those of the singular region around the star Rho Ophiuchi. Here are to be seen somber lanes and patches, apparently forming a connected system which covers an immense space, and which their discoverer thinks may constitute a "dark nebula." This seems at first a startling suggestion; but, after all, why should their not be dark nebulæ as well as visible ones? In truth, it has troubled some astronomers to explain the luminosity of the bright nebulæ, since it is not to be supposed that matter in so diffuse a state can be incandescent through heat, and phosphorescent light is in itself a mystery. The supposition is also in accord with what we know of the existence of dark solid bodies in space. Many bright stars are accompanied by obscure companions, sometimes as massive as themselves; the planets are non-luminous; the same is true of meteors before they plunge into the atmosphere and become heated by friction; and many plausible reasons have been found for believing that space contains as many obscure as shining bodies of great size. It is not so difficult, after all, then, to believe that there are immense collections of shadowy gases and meteoric dust whose presence is only manifested when they intercept the light coming from shining bodies behind them.

This would account for the apparent extinguishment of light in open space, which is indicated by the falling off in relative number of telescopic stars below the tenth magnitude. Even as things are, the amount of light coming to us from stars too faint to be seen with the naked eye is so great that the statement of it generally surprises persons who are unfamiliar with the inner facts of astronomy. It has been calculated that on a clear night the total starlight from the entire celestial sphere amounts to one-sixtieth of the light of the full moon; but of this less than one-twenty-fifth is due to stars separately distinguished by the eye. If there were no obscuring medium in space, it is probable that the amount of starlight would be noticeably and perhaps enormously increased.

But while it seems certain that some of the obscure spots in the Milky Way are due to the presence of "dark nebulæ," or concealing veils of one kind or another, it is equally certain that there are many which are true apertures, however they may have been formed, and by whatever forces they may be maintained. These, then, are veritable windows of the Galaxy, and when looking out of them one is face to face with the great mystery of infinite space. *There* the known universe visibly ends, but manifestly space itself does not end there. It is not within the power of thought to conceive an end to space, for the instant we think of a terminal point or line the mind leaps forward to the *beyond*. There must be space outside as well as inside. Eternity of time and infinity of space are ideas that the intellect cannot fully grasp, but neither can it grasp the idea of a limitation to either space or time. The metaphysical conceptions of hypergeometry, or fourth-dimensional space, do not aid us.

Having, then, discovered that the universe is a thing *contained* in something indefinitely greater than itself; having looked out of its windows and found only the gloom of starless night outside — what conclusions are we to draw concerning the beyond? It *seems* as empty as a vacuum, but is it really so? If it be, then our universe is a single atom astray in the infinite; it is the only island in an ocean without shores;

it is the one oasis in an illimitable desert. Then the Milky Way, with its wide-flung garland of stars, is afloat like a tiny smoke-wreath amid a horror of immeasurable vacancy, or it is an evanescent and solitary ring of sparkling froth cast up for a moment on the viewless billows of immensity. From such conclusions the mind instinctively shrinks. It prefers to think that there is *something* beyond, though we cannot see it. Even the universe could not bear to be alone — a Crusoe lost in the Cosmos! As the inhabitants of the most elegant château, with its gardens, parks, and crowds of attendants, would die of loneliness if they did not know that they have neighbors, though not seen, and that a living world of indefinite extent surrounds them, so we, when we perceive that the universe has limits, wish to feel that it is not solitary; that beyond the hedges and the hills there are other centers of life and activity. Could anything be more terrible than the thought of an *isolated universe?* The greater the being, the greater the aversion to seclusion. Only the infinite satisfies; in that alone the mind finds rest.

We are driven, then, to believe that the universal night which envelopes us is not tenantless; that as we stare out of the star-framed windows of the Galaxy and see nothing but uniform blackness, the fault is with our eyes or is due to an obscuring medium. Since *our* universe is limited in extent, there must be *other* universes beyond it on all sides. Perhaps if we could carry our telescopes to the verge of the great "Coal-sack" near the "Cross," being then on the frontier of our starry system, we could discern, sparkling afar off in the vast night, some of the outer galaxies. They may be grander than ours, just as many of the suns surrounding us are immensely greater than ours. If we could take our stand somewhere in the midst of immensity and, with vision of infinite reach, look about us, we should perhaps see a countless number of stellar systems, amid which ours would be unnoticeable, like a single star among the multitude glittering in the terrestrial sky on a clear night. Some might be in the form of a wreath, like our own; some might be globular, like the great star-clusters in Hercules and Centaurus; some might be glittering circles, or disks, or rings within rings. If we could enter them we should probably find a vast variety of composition, including elements unknown to terrestrial chemistry; for while the *visible* universe appears to contain few if any substances not existing on the earth or in the sun, we have no warrant to assume that others may not exist in infinite space.

And how as to gravitation? We do not *know* that gravitation acts beyond the visible universe, but it is reasonable to suppose that it does. At any rate, if we let go *its* sustaining hand we are lost, and can only wander hopelessly in our speculations, like children astray. If the empire of gravitation is infinite, then the various outer systems must have *some*, though measuring by our standards an imperceptible, attractive influence upon each other, for gravitation never lets go its hold, however great the space over which it is required to act. Just as the stars about us are all in motion, so the starry systems beyond our sight may be in motion, and our system as a whole may be moving in concert with them. If this be so, then after interminable ages the aspect of the entire system of systems must change, its various members assuming new positions with respect to one another. In the course of time we may even suppose that our universe will approach relatively close to one of the others; and then, if men are yet living on the earth, they may glimpse through the openings which reveal nothing to us now, the lights of another nearing star system, like the signals of a strange squadron, bringing them the assurance (which can be but an inference at present) that the ocean of space has other argosies venturing on its limitless expanse.

There remains the question of the luminiferous ether by whose agency the waves of light are borne through space. The ether is as mysterious as gravitation. With regard to ether we only infer its existence from the effects which we ascribe to it. Evidently the ether must extend as far as the most distant visible stars. But does it continue on indefinitely in outer space? If it does, then the invisibility of the other systems must be due to their distance diminishing the quantity of light that comes from them below the limit of perceptibility, or to the interposition of absorbing media; if it does not, then the reason why we cannot see them is owing to the absence of a means of conveyance for the light waves, as the lack of an interplanetary atmosphere prevents us from hearing the thunder of sun-spots. (It is interesting to recall that Mr. Edison was once credited with the intention to construct a gigantic microphone which should render the roar of sun-spots audible by transforming the electric vibrations into sound-waves). On this supposition each starry system would be enveloped in its own globule of ether, and no light could cross from one to another. But the probability is that both the ether and gravitation are ubiquitous, and that all the stellar systems are immersed in the former like clouds of phosphorescent organisms in the sea.

So astronomy carries the mind from height to greater height. Men were long in accepting the proofs of the relative insignificance of the earth; they were more quickly convinced of the comparative littleness of the solar system; and now the evidence assails their reason that what they had regarded as *the* universe is only one mote gleaming in the sunbeams of Infinity.

Star-Clouds, Star-Clusters, and Star-Streams

*I*n the preceding chapter we have seen something of the strangely complicated structure of the Galaxy, or Milky Way. We now proceed to study more comprehensively that garlanded "Pathway of the Gods."

Judged by the eye alone, the Milky Way is one of the most delicately beautiful phenomena in the entire realm of nature — a shimmer of silvery gauze stretched across the sky; but studied in the light of its revelations, it is the most stupendous object presented to human ken. Let us consider, first, its appearance to ordinary vision. Its apparent position in the sky shifts according to the season. On a serene, cloudless summer evening, in the absence of the moon, whose light obscures it, one sees the Galaxy spanning the heavens from north to southeast of the zenith like a phosphorescent arch. In early spring it forms a similar but, upon the whole, less brilliant arch west of the zenith. Between spring and summer it lies like a long, faint, twilight band along the northern horizon. At the beginning of winter it again forms an arch, this time spanning the sky from east to west, a little north of the zenith. These are its positions as viewed from the mean latitude of the United States. Even the beginner in star-gazing does not have to watch it throughout the year in order to be convinced that it is, in reality, a great circle, extending entirely around the celestial sphere. We appear to be situated near its center, but its periphery is evidently far away in the depths of space.

Although to the casual observer it seems but a delicate scarf of light, brighter in some places than in others, but hazy and indefinite at the best, such is not its appearance to those who study it with care. They perceive that it is an organic whole, though marvelously complex in detail. The telescope shows that it consists of stars too faint and small through excess of distance to be separately visible. Of the hundred million suns which some estimates have fixed as the probable population of the starry universe, the vast majority (at least thirty to one) are included in this strange belt of misty light. But they are not uniformly distributed in it; on the contrary, they are arrayed in clusters, knots, bunches, clouds, and streams. The appearance is somewhat as if the Galaxy consisted of innumerable swarms of silver-winged bees, more or less intermixed, some massed together, some crossing the paths of others, but all governed by a single purpose which leads them to encircle the region of space in which we are situated.

From the beginning of the systematic study of the heavens, the fact has been recognized that the form of the Milky Way denotes the scheme of the sidereal system. At first it was thought that the shape of the system was that of a vast round disk, flat like a cheese, and filled with stars, our sun and his relatively few neighbors being placed near the center. According to this view, the galactic belt was an effect of perspective; for when looking in the direction of the plane of the disk, the eye ranged through an immense extension of stars which blended into a glimmering blur, surrounding us like a ring; while when looking out from the sides of the disk we saw but few stars, and in those directions the heavens appeared relatively blank. Finally it was recognized that this theory did not correspond with the observed appearances, and it became evident that the Milky Way was not a mere effect of perspective, but an actual band of enormously distant stars, forming a circle about the sphere, the central opening of the ring (containing many scattered stars) being many times broader than the width of the ring itself. Our sun is one of the scattered stars in the central opening.

As already remarked, the ring of the Galaxy is very irregular, and in places it is partly broken. With its sinuous outline, its pendant sprays, its graceful and accordant curves, its bunching of masses, its occasional interstices, and the manifest order of a general plan governing the jumble of its details, it bears a remarkable resemblance to a garland — a fact which appears the more wonderful when we recall its composition. That an elm tree should trace the lines of beauty with its leafy and pendulous branches does not surprise us; but we can only gaze with growing amazement when we behold *a hundred million suns imitating the form of a chaplet!* And then we have to remember that this form furnishes the ground-plan of the universe.

As an indication of the extraordinary speculations to which the mystery of the Milky Way has given rise, a theory recently (1909) proposed by Prof. George C. Comstock may be mentioned. Starting with the data (first) that the number of stars increases as the Milky Way is approached, and reaches a maximum in its plane, while on the other hand the number of nebulæ is greatest outside the Milky Way and increases with distance from it, and (second) that the Milky Way, although a complete ring, is broad and diffuse on one side through one-half its course — that half alone containing nebulæ — and relatively narrow and well defined on the opposite side, the author of this singular speculation avers that these facts can best be explained by supposing that the invisible universe consists of two interpenetrating parts, one of which is a chaos of indefinite extent, strewn with stars and nebulous dust, and the other a long, broad but comparatively thin cluster of stars, including the sun as one of its central members. This flat star-cluster is conceived to be moving edgewise through the chaos, and, according to Professor Comstock, it acts after the manner of a snow-plow sweeping away the cosmic dust and piling it on either hand above and below the plane of the moving cluster. It thus forms a transparent rift, through which we see farther and command a view of more stars than through the intensified dust-clouds on either hand. This rift is the Milky Way. The dust thrown aside toward the poles of the Milky Way is the substance of the nebulæ which abound there. Ahead, where the front of the star-plow is clearing the way, the chaos is nearer at hand, and consequently there the rift subtends a broader angle, and is filled with primordial dust, which, having been annexed by the vanguard of the star-swarm, forms the nebulæ seen only in that part of the Milky Way. But behind, the rift appears narrow because there we look farther away between dust-clouds produced ages ago by the front of the plow, and no scattered dust remains in that part of the rift.

In quoting an outline of this strikingly original theory the present writer should not be understood as assenting to it. That it appears bizarre is not, in itself, a reason for rejecting it, when we are dealing with so problematical and enigmatical a subject as the Milky Way; but the serious objection is that the theory does not sufficiently accord with the observed phenomena. There is too much evidence that the Milky Way is an organic system, however fantastic its form, to permit the belief that it can only be a rift in chaotic clouds. As with every organism, we find that its parts are more or less clearly repeated in its ensemble. Among all the strange things that the Milky Way contains there is nothing so extraordinary as itself. Every astronomer must many times have found himself marveling at it in those comparatively rare nights when it shows all its beauty and all its strangeness. In its great broken rifts, divisions, and spirals are found the gigantic prototypes of similar forms in its star-clouds and clusters. As we have said, it determines the general shape of the whole sidereal system. Some of the brightest stars in the sky appear to hang like jewels suspended at the ends of tassels dropped from the Galaxy. Among these pendants are the Pleiades and the Hyades. Orion, too, the "Mighty Hunter," is caught in "a loop of light" thrown out from it. The majority of the great first-magnitude stars seem related to it, as if they formed an inner ring inclined at an angle of some twenty degrees to its plane. Many of the long curves that set off from it on both sides are accompanied by corresponding curves of lucid stars. In a word, it offers every appearance of structural connection with the entire starry system. That the universe should have assumed the form of a wreath is certainly a matter for astonishment; but it would have been still more astonishing if it had been a cube, a rhomboid, or a dodecahedron, for then we should have had to suppose that something resembling the forces that shape crystals had acted upon the stars, and the difficulty of explaining the universe by the laws of gravitation would have been increased.

From the Milky Way as a whole we pass to the vast clouds, swarms, and clusters of stars of which it is made up. It may be, as some astronomers hold, that most of the galactic stars are much smaller than the sun, so that their faintness is not due entirely to the effect of distance. Still, their intrinsic brilliance attests their solar character, and considering their remoteness, which has been estimated at not less than ten thousand to twenty thousand light-years (a light-year is equal to nearly six thousand thousand million miles) their actual masses cannot be extremely small. The minutest of them are entitled to be regarded as real suns, and they vary enormously in magnitude. The effects of their attractions upon one another can only be inferred from their clustering, because their relative movements are not apparent on account of the brevity of the observations that we can make. But imagine a being for whom a million years would be but as a flitting moment; to him the Milky Way would appear in a state of ceaseless agitation — swirling with "a fury of whirlpool motion."

The cloudlike aspect of large parts of the Galaxy must always have attracted attention, even from naked-eye observers, but the true star-clouds were first satisfactorily represented in Barnard's photographs. The resemblance to actual clouds is often startling. Some are close-packed and dense, like cumuli; some are wispy or mottled, like cirri. The rifts and modulations, as well as the general outlines, are the same as those of clouds of vapor or dust, and one notices also the characteristic thinning out at the edges. But we must beware of supposing that the component suns are thickly crowded as the particles forming an ordinary cloud. They *look*, indeed, as if they were matted together, because of the irradiation of light, but in reality millions and billions of miles separate each star from its neighbors. Nevertheless they form real assemblages, whose members are far more closely related to one another than is our sun to the stars around him, and if we were in the Milky Way the aspect of the nocturnal sky would be marvelously different from its present appearance.

Stellar clouds are characteristic of the Galaxy and are not found beyond its borders, except in the "Magellanic Clouds" of the southern hemisphere, which resemble detached portions of the Milky Way. These singular objects form as striking a peculiarity of the austral heavens as does the great "Coal-sack" described in Chapter 1. But it is their isolation that makes them so remarkable, for their composition is essentially galactic, and if they were included within its boundaries they would not appear more wonderful than many other parts of the Milky Way. Placed where they are, they look like masses fallen from the great stellar arch. They are full of nebulæ and star-clusters, and show striking evidences of spiral movement.

Star-swarms, which are also characteristic features of the Galaxy, differ from star-clouds very much in the way that their name would imply — *i.e.,* their component stars are so arranged, even when they are countless in number, that the idea of an exceedingly numerous assemblage rather than that of a cloud is impressed on the observer's mind. In a star-swarm the separate members are distinguishable because they are either larger or nearer than the stars composing a "cloud." A splendid example of a true star-swarm is furnished by Chi Persei, in that part of the Milky Way which runs between the constellations Perseus and Cassiopeia. This swarm is much coarser than many others, and can be seen by the naked eye. In a small telescope it appears double, as if the suns composing it had divided into two parties which keep on their way side by side, with some commingling of their members where the skirts of the two companies come in contact.

Smaller than either star-clouds or star-swarms, and differing from both in their organization, are star-clusters. These, unlike the others, are found outside as well as inside the Milky Way, although they are more numerous inside its boundaries than elsewhere. The term star-cluster is sometimes applied, though improperly, to assemblages which are rather groups, such, for instance, as the Pleiades. In their most characteristic aspect star-clusters are of a globular shape — globes of suns! A famous example of a globular star-cluster, but one not included in the Milky Way, is the "Great Cluster in Hercules." This is barely visible to the naked eye, but a small telescope shows its character, and in a large one it presents a marvelous spectacle. Photographs of such clusters are, perhaps, less effective than those of star-clouds, because the central condensation of stars in them is so great that their light becomes blended in an indistinguishable blur. The beautiful effect of the incessant play of infinitesimal rays over the apparently compact surface of the cluster, as if it were a globe of the finest frosted silver shining in an electric beam, is also lost in a photograph. Still, even to the eye looking directly at the cluster through a powerful telescope, the central part of the wonderful congregation seems almost a solid mass in which the stars are packed like the ice crystals in a snowball.

The same question rises to the lips of every observer: How can they possibly have been brought into such a situation? The marvel does not grow less when we know that, instead of being closely compacted, the stars of the cluster are probably separated by millions of miles; for we know that their distances apart are slight as compared with their remoteness from the Earth. Sir William Herschel estimated their number to be about fourteen thousand, but in fact they are uncountable. If we could view them from a point just within the edge of the assemblage, they would offer the appearance of a hollow hemisphere emblazoned with stars of astonishing brilliancy; the near-by ones unparalleled in splendor by any celestial object known to us, while the more distant ones would resemble ordinary stars. An inhabitant of the cluster would not know, except by a process of ratiocination, that he was dwelling in a globular assemblage of suns; only from a point far outside would their spherical arrangement become evident to the eye. Imagine fourteen-thousand fire-balloons with an approach to regularity in a spherical space — say, ten miles in diameter; there would be an average of less than thirty in every cubic mile, and it would be necessary to go to a considerable distance in order to see them as a globular aggregation; yet from a point sufficiently far away they would blend into a glowing ball.

Photographs show even better than the best telescopic views that the great cluster is surrounded with a multitude of dispersed stars, suggestively arrayed in more or less curving lines, which radiate from the principle mass, with which their connection is manifest. These stars, situated outside the central sphere, look somewhat like vagrant bees buzzing round a dense swarm where the queen bee is sitting. Yet while there is so much to suggest the operation of central forces, bringing and keeping the members of the cluster together, the attentive observer is also impressed with the idea that the whole wonderful phenomenon may be *the result of explosion.* As soon as this thought seizes the mind, confirmation of it seems to be found in the appearance of the outlying stars, which could be as readily explained by the supposition that they have been blown apart as that they have flocked together toward a center. The probable fact that the stars constituting the cluster are very much smaller than our sun might be regarded as favoring the hypothesis of an explosion. Of their real size we know nothing, but, on the basis of an uncertain estimate of their parallax, it has been calculated that they may average forty-five thousand miles in diameter — something more than half the diameter of the planet Jupiter. Assuming the same mean density, fourteen thousand such stars might have been formed by the explosion of a body about twice the size of the sun. This recalls the theory of Olbers, which has never been altogether abandoned or disproved, that the Asteroids were formed by the explosion of a planet circulating between the orbits of Mars and Jupiter. The Asteroids, whatever their manner of origin, form a ring around the sun; but, of course, the explosion of a great independent body, not originally revolving about a superior center of gravitational force, would not result in the formation of a ring of small bodies, but rather of a dispersed mass of them. But back of any speculation of this kind lies the problem, at present insoluble: How could the explosion be produced? (See the question of explosions in Chapters 6 and 14).

Then, on the other hand, we have the observation of Herschel, since abundantly confirmed, that space is unusually vacant in the immediate neighborhood of condensed star-clusters and nebulæ, which, as far as it goes, might be taken as an indication that the assembled stars had been drawn together by their mutual attractions, and that the tendency to aggregation is still bringing new members toward the cluster. But in that case there must have been an original condensation of stars at that point in space. This could probably have been produced by the coagulation of a great nebula into stellar nuclei, a process which seems now to be taking place in the Orion Nebula.

A yet more remarkable globular star-cluster exists in the southern hemisphere, Omega Centauri. In this case the central condensation of stars presents an almost uniform blaze of light. Like the Hercules cluster, that in Centaurus is surrounded with stars scattered over a broad field and showing an appearance of radial arrangement. In fact, except for its greater richness, Omega Centauri is an exact duplicate of its northern rival. Each appears to an imaginative spectator as a veritable "city of suns." Mathematics shrinks from the task of disentangling the maze of motions in such an assemblage. It would seem that the chance of collisions is not to be neglected, and this idea finds a certain degree of confirmation in the appearance of "temporary stars" which have more than once blazed out in, or close by, globular star-clusters.

This leads up to the notable fact, first established by Professor Bailey a few years ago, that such clusters are populous with variable stars. Omega Centauri and the Hercules cluster are especially remarkable in this respect. The variables found in them are all of short period and the changes of light show a noteworthy tendency to uniformity. The first thought is that these phenomena must be due to collisions among the crowded stars, but, if so, the encounters cannot be between the stars themselves, but probably between stars and meteor swarms revolving around them. Such periodic collisions might go on for ages without the meteors being exhausted by incorporation with the stars. This explanation appears all the more probable because one would naturally expect that flocks of meteors would abound in a close aggregation of stars. It is also consistent with Perrine's discovery — that the globular star clusters are powdered with minute stars strewn thickly among the brighter ones.

In speaking of Professor Comstock's extraordinary theory of the Milky Way, the fact was mentioned that, broadly speaking, the nebulæ are less numerous in the galactic belt than in the comparatively open spaces on either side of it, but that they are, nevertheless, abundant in the broader half of the Milky Way which he designates as the front of the gigantic "plow" supposed to be forcing its way through the enveloping chaos. In and around the Sagittarius region the intermingling of nebulæ and galactic star clouds and clusters is particularly remarkable. That there is a causal connection no thoughtful person can doubt. We are unable to get away from the evidence that a nebula is like a seed-ground from which stars spring forth; or we may say that nebulæ resemble clouds in whose bosom raindrops are forming. The wonderful aspect of the admixtures of nebulæ and star-clusters in Sagittarius has been described in Chapter 1. We now come to a still more extraordinary phenomenon of this kind — the Pleiades nebulæ.

The group of the Pleiades, although lying outside the main course of the Galaxy, is connected with it by a faint loop, and is the scene of the most remarkable association of stars and nebulous matter known in the visible universe. The naked eye is unaware of the existence of nebulæ in the Pleiades, or, at the best, merely suspects that there is something of the kind there; and even the most powerful telescopes are far from revealing the full wonder of the spectacle; but in photographs which have been exposed for many hours consecutively, in order to accumulate the impression of the actinic rays, the revelation is stunning. The principle stars are seen surrounded by, and, as it were, *drowned in*, dense nebulous clouds of an unparalleled kind. The forms assumed by these clouds seem at first sight inexplicable. They look like fleeces, or perhaps more like splashes and daubs of luminous paint dashed carelessly from a brush. But closer inspection shows that they are, to a large extent, *woven* out of innumerable threads of filmy texture, and there are many indications of spiral tendencies. Each of the bright stars of the group — Alcyone, Merope, Maia, Electra, Taygeta, Atlas — is the focus of a dense fog (totally invisible, remember, alike to the naked eye and to the telescope), and these particular stars are veiled from sight behind the strange mists. Running in all directions across the relatively open spaces are nebulous wisps and streaks of the most curious forms. On some of the nebular lines, which are either straight throughout, or if they change direction do so at an angle, little stars are strung like beads. In one case seven or eight stars are thus aligned, and, as if to emphasize their dependence upon the chain which connects them, when it makes a slight bend the file of stars turns the same way. Many other star rows in the group suggest by their arrangement that they, too, were once strung upon similar threads which have now disappeared, leaving the stars spaced along their ancient tracks. We seem forced to the conclusion that there was a time when the Pleiades were embedded in a vast nebula resembling that of Orion, and that the cloud has now become so rare by gradual condensation into stars that the merest trace of it remains, and this would probably have escaped detection but for the remarkable actinic power of the radiant matter of which it consists. The richness of many of these faint nebulous masses in ultra-violet radiations, which are those that specifically affect the photographic plate, is the cause of the marvelous revelatory power of celestial photography. So the veritable unseen universe, as distinguished from the "unseen universe" of metaphysical speculation, is shown to us.

A different kind of association between stars and nebulæ is shown in some surprising photographic objects in the constellation Cygnus, where long, wispy nebulæ, billions of miles in length, some of them looking like tresses streaming in a breeze, lie amid fields of stars which seem related to them. But the relation is of a most singular kind, for notwithstanding the delicate structure of the long nebulæ they appear to act as barriers, causing the stars to heap themselves on one side. The stars are two, three, or four times as numerous on one side of the nebulæ as on the other. These nebulæ, as far as appearance goes, might be likened to rail fences, or thin hedges, against which the wind is driving drifts of powdery snow, which, while scattered plentifully all around, tends to bank itself on the leeward side of the obstruction. The imagination is at a loss to account for these extraordinary phenomena; yet there they are, faithfully giving us their images whenever the photographic plate is exposed to their radiations.

Thus the more we see of the universe with improved methods of observation, and the more we invent aids to human senses, each enabling us to penetrate a little deeper into the unseen, the greater becomes the mystery. The telescope carried us far, photography is carrying us still farther;

but what as yet unimagined instrument will take us to the bottom, the top, and the end? And then, what hitherto untried power of thought will enable us to comprehend the meaning of it all?

Stellar Migrations

To the untrained eye the stars and the planets are not distinguishable. It is customary to call them all alike "stars." But since the planets more or less rapidly change their places in the sky, in consequence of their revolution about the sun, while the stars proper seem to remain always in the same relative positions, the latter are spoken of as "fixed stars." In the beginnings of astronomy it was not known that the "fixed stars" had any motion independent of their apparent annual revolution with the whole sky about the earth as a seeming center. Now, however, we know that the term "fixed stars" is paradoxical, for there is not a single really fixed object in the whole celestial sphere. The apparent fixity in the positions of the stars is due to their immense distance, combined with the shortness of the time during which we are able to observe them. It is like viewing the plume of smoke issuing from a steamer, hull down, at sea: if one does not continue to watch it for a long time it appears to be motionless, although in reality it may be traveling at great speed across the line of sight. Even the planets seem fixed in position if one watches them for a single night only, and the more distant ones do not sensibly change their places, except after many nights of observation. Neptune, for instance, moves but little more than two degrees in the course of an entire year, and in a month its change of place is only about one-third of the diameter of the full moon.

Yet, fixed as they seem, the stars are actually moving with a speed in comparison with which, in some cases, the planets might almost be said to stand fast in their tracks. Jupiter's speed in his orbit is about eight miles per second, Neptune's is less than three and one-half miles, and the earth's is about eighteen and one-half miles; while there are "fixed stars" which move two hundred or three hundred miles per second. They do not all, however, move with so great a velocity, for some appear to travel no faster than the planets. But in all cases, notwithstanding their real speed, long-continued and exceedingly careful observations are required to demonstrate that they are moving at all. No more overwhelming impression of the frightful depths of space in which the stars are buried can be obtained than by reflecting upon the fact that a star whose actual motion across the line of sight amounts to two hundred miles per second does not change its apparent place in the sky, in the course of a thousand years, sufficiently to be noticed by the casual observer of the heavens!

There is one vast difference between the motions of the stars and those of the planets to which attention should be at once called: the planets, being under the control of a central force emanating from their immediate master, the sun, all move in the same direction and in orbits concentric about the sun; the stars, on the other hand, move in every conceivable direction and have no apparent center of motion, for all efforts to discover such a center have failed. At one time, when theology had finally to accept the facts of science, a grandiose conception arose in some pious minds, according to which the Throne of God was situated at the exact center of His Creation, and, seated there, He watched the magnificent spectacle of the starry systems obediently revolving around Him. Astronomical discoveries and speculations seemed for a time to afford some warrant for this view, which was, moreover, an acceptable substitute for the abandoned geocentric theory in minds that could only conceive of God as a superhuman artificer, constantly admiring his own work. No longer ago than the middle of the nineteenth century a German astronomer, Maedler, believed that he had actually found the location of the center about which the stellar universe revolved. He placed it in the group of the Pleiades, and upon his authority an extraordinary imaginative picture was sometimes drawn of the star Alcyone, the brightest of the Pleiades, as the very seat of the Almighty. This idea even seemed to gain a kind of traditional support from the mystic significance, without known historical origin, which has for many ages, and among widely separated peoples, been attached to the remarkable group of which Alcyone is the chief. But since Maedler's time it has been demonstrated that the Pleiades cannot be the center of revolution of the universe, and, as already remarked, all attempts to find or fix such a center have proved abortive. Yet so powerful was the hold that the theory took upon the popular imagination, that even today astronomers are often asked if Alcyone is not the probable site of "Jerusalem the Golden."

If there were a discoverable center of predominant gravitative power, to which the motions of all the stars could be referred, those motions would appear less mysterious, and we should then be able to conclude that the universe was, as a whole, a prototype of the subsidiary systems of which it is composed. We should look simply to the law of gravitation for an explanation, and, naturally, the center would be placed within the opening enclosed by the Milky Way. If it were there the Milky Way itself should exhibit signs of revolution about it, like a wheel turning upon its hub. No theory of the star motions as a whole could stand which failed to take account of the Milky Way as the basis of all. But the very form of that divided wreath of stars forbids the assumption of its revolution about a center. Even if it could be conceived as a wheel having no material center it would not have the form which it actually presents. As was shown in Chapter 2, there is abundant evidence of motion in the Milky Way; but it is not motion of the system as a whole, but motion affecting its separate parts. Instead of all moving one way, the galactic stars, as far as their movements can be inferred, are governed by local influences and conditions. They appear to travel crosswise and in contrary directions, and perhaps they eddy around foci where great numbers have assembled; but of a universal revolution involving the entire mass we have no evidence.

Most of our knowledge of star motions, called "proper motions," relates to individual stars and to a few groups which happen to be so near that the effects of their movements are measurable. In some cases the motion is so rapid (not in appearance, but in reality) that the chief difficulty is to imagine how it can have been imparted, and what will eventually become of the "runaways." Without a collision, or a series of very close approaches to great gravitational centers, a star traveling through space at the rate of two hundred or three hundred miles per second could not be arrested or turned into an orbit which would keep it forever flying within the limits of the visible universe. A famous example of these speeding stars is "1830 Groombridge," a star of only the sixth magnitude, and consequently just visible to the naked eye, whose motion across the line of sight is so rapid that it moves upon the face of the sky a distance equal to the apparent diameter of the moon every 280 years. The distance of this star is at least 200,000,000,000,000 miles, and may be two or three times greater, so that its actual speed cannot be less than two hundred, and may be as much as four hundred, miles per second. It could be turned into a new course by a close approach to a great sun, but it could only be stopped by collision, head-on, with a body of enormous mass. Barring such accidents it must, as far as we can see, keep on until it has traversed our stellar system, whence in may escape and pass out into space beyond, to join, perhaps, one of those other universes of which we have spoken. Arcturus, one of the greatest suns in the universe, is also a runaway, whose speed of flight has been estimated all the way from fifty to two hundred miles per second. Arcturus, we have every reason to believe, possesses hundreds of times the mass of our sun – think, then, of the prodigious momentum that its motion implies! Sirius moves more moderately, its motion across the line of sight amounting to only ten miles per second, but it is at the same time approaching the sun at about the same speed, its actual velocity in space being the resultant of the two displacements.

What has been said about the motion of Sirius brings us to another aspect of this subject. The fact is, that in every case of stellar motion the displacement that we observe represents only a part of the actual movement of the star concerned. There are stars whose motion carries them straight toward or straight away from the earth, and such stars, of course, show no cross motion. But the vast majority are traveling in paths inclined from a perpendicular to our line of sight. Taken as a whole, the stars may be said to be flying about like the molecules in a mass of gas. The discovery of the radial component in the movements of the stars is due to the spectroscope. If a star is approaching, its spectral lines are shifted toward the violet end of the spectrum by an amount depending upon the velocity of approach; if it is receding, the lines are correspondingly shifted toward the red end. Spectroscopic observation, then, combined with micrometric measurements of the cross motion, enables us to detect the real movement of the star in space. Sometimes it happens that a star's radial movement is periodically reversed; first it approaches, and then it recedes. This indicates that it is revolving around a near-by companion, which is often invisible, and superposed upon this motion is that of the two stars concerned, which together may be approaching or receding or traveling across the line of sight. Thus the complications involved in the stellar motions are often exceedingly great and puzzling.

Yet another source of complication exists in the movement of our own star, the sun. There is no more difficult problem in astronomy than that of disentangling the effects of the solar motion from those of the motions of the other stars. But the problem, difficult as it is, has been solved, and upon its solution depends our knowledge of the speed and direction of the movement of the solar system through space, for of course the sun carries its planets with it. One element of the solution is found in the fact that, as a result of perspective, the stars toward which we are going appear to move apart toward all points of the compass, while those behind appear to close up together. Then the spectroscopic principle already mentioned is invoked for studying the shift of the lines, which is toward the violet in the stars ahead of us and toward the red in those that we are leaving behind. Of course the effects of the independent motions of the stars must be carefully excluded. The result of the studies devoted to this subject is to show that we are traveling at a speed of twelve to fifteen miles per second in a northerly direction, toward the border of the constellations Hercules and Lyra. A curious fact is that the more recent estimates show that the direction is not very much out of a straight line drawn from the sun to the star Vega, one of the most magnificent suns in the heavens. But it should not be inferred from this that Vega is drawing us on; it is too distant for its gravitation to have such an effect.

Many unaccustomed thoughts are suggested by this mighty voyage of the solar system. Whence have we come, and whither do we go? Every year of our lives we advance at least 375,000,000 miles. Since the traditional time of Adam the sun has led his planets through the wastes of space no less than 225,000,000,000 miles, or more than 2400 times the distance that separates him from the earth. Go back in imagination to the geologic ages, and try to comprehend the distance over which the earth has flown. Where was our little planet when it emerged out of the clouds of chaos? Where was the sun when his "thunder march" began? What strange constellations shone down upon our globe when its masters of life were the monstrous beasts of the "Age of Reptiles?" A million years is not much of a span of time in geologic reckoning, yet a million years ago the earth was farther from its present place in space than any of the stars with a measurable parallax are now. It was more than seven times as far as Sirius, nearly fourteen times as far as Alpha Centauri, three times as far as Vega, and twice as far as Arcturus. But some geologists demand two hundred, three hundred, even one thousand million years to enable them to account for the evolutionary development of the earth and its inhabitants. In a thousand million years the earth would have traveled farther than from the remotest conceivable depths of the Milky Way!

Other curious reflections arise when we think of the form of the earth's track as it follows the lead of the sun, in a journey which has neither known beginning nor conceivable end. There are probably many minds which have found a kind of consolation in the thought that every year the globe returns to the same place, on the same side of the sun. This idea may have an occult connection with our traditional regard for anniversaries. When that period of the year returns at which any great event in our lives has occurred we have the feeling that the earth, in its annual round, has, in a manner, brought us back to the scene of that event. We think of the earth's orbit as a well-worn path which we traverse many times in the course of a lifetime. It seems familiar to us, and we grow to have a sort of attachment to it. The sun we are accustomed to regard as a fixed center in space, like the mill or pump around which the harnessed patient mule makes his endless circuits. But the real fact is that the earth never returns to the place in space where it has once quitted. In consequence of the motion of the sun carrying the earth and the other planets along, the track pursued by our globe is a vast spiral in space continually developing and never returning upon its course. It is probable that the tracks of the sun and the others stars are also irregular, and possibly spiral, although, as far as can be at present determined, they appear to be practically straight. Every star, wherever it may be situated, is attracted by its fellow-stars from many sides at once, and although the force is minimized by distance, yet in the course of many ages its effects must become manifest.

Looked at from another side, is there not something immensely stimulating and pleasing to the imagination in the idea of so stupendous a journey, which makes all of us the greatest of travelers? In the course of a long life a man is transported through space thirty thousand million miles; Halley's Comet does not travel one-quarter as far in making one of its immense circuits. And there are adventures on this voyage of which we are just beginning to learn to take account. Space is full of strange things, and the earth must encounter some of them as it advances through the unknown. Many singular speculations have been indulged in by astronomers concerning the possible effects upon the earth of the varying state of the space that it traverses. Even the alternation of hot and glacial periods has sometimes been ascribed to this source. When tropical life flourished around the poles, as the remains in the rocks assure us, the needed high temperature may, it has been thought, have been derived from the presence of the earth in a warm region of space. Then, too, there is a certain interest for us in the thought of what our familiar planet has passed through. We cannot but admire it for its long journeying as we admire the traveler who comes to us from remote and unexplored lands, or as we gaze with a glow of interest upon the first locomotive that has crossed a continent, or a ship that has visited the Arctic or Antarctic regions. If we may trust the indications of the present course, the earth, piloted by the sun, has come from the Milky Way in the far south and may eventually rejoin that mighty band of stars in the far north.

While the stars in general appear to travel independently of one another, except when they are combined in binary or trinary systems, there are notable exceptions to this rule. In some quarters of the sky we behold veritable migrations of entire groups of stars whose members are too widely separated to show any indications of revolution about a common center of gravity. This leads us back again to the wonderful group of the Pleiades. All of the principle stars composing that group are traveling in virtually parallel lines. Whatever force set them going evidently acted upon all alike. This might be explained by the assumption that when the original projective force acted upon them they were more closely united than they are at present, and that in drifting apart they have not lost the impulse of the primal motion. Or it may be supposed that they are carried along by some current in space, although it would be exceedingly difficult, in the present state of our knowledge, to explain the nature of such a current. Yet the theory of a current has been proposed. As to an attractive center around which they might revolve, none has been found. Another instance of similar "star-drift" is furnished by five of the seven stars constituting the figure of the "Great Dipper." In this case the stars concerned are separated very widely, the two extreme ones by not less than fifteen degrees, so that the idea of a common motion would never have been suggested by their aspect in the sky; and the case becomes the more remarkable from the fact that among and between them there are other stars, some of the same magnitude, which do not share their motion, but are traveling in other directions. Still other examples of the same phe-nomenon are found in other parts of the sky. Of course, in the case of compact star-clusters, it is assumed that all the members share a like motion of translation through space, and the same is probably true of dense star-swarms and star-clouds.

The whole question of star-drift has lately assumed a new phase, in consequence of the investigations of Kapteyn, Dyson, and Eddington on the "systematic motions of the stars." This research will, it is hoped, lead to an understanding of the general law governing the movements of the whole body of stars constituting the visible universe. Taking about eleven hundred stars whose proper motions have been ascertained with an approach to certainty, and which are distributed in all parts of the sky, it has been shown that there exists an apparent double drift, in two independent streams, moving in different and nearly opposed directions. The apex of the motion of what is called "Stream I" is situated, according to Professor Kapteyn, in right ascension 85°, declination south 11°, which places it just south of the constellation Orion; while the apex of "Stream II" is in right ascension 260°, declination south 48°, placing it in the constellation Ara, south of Scorpio. The two apices differ very nearly 180° in right ascension and about 120° in declination. The discovery of these vast star-streams, if they really exist, is one of the most extraordinary in modern astronomy. It offers the correlation of stellar movements needed as the basis of a theory of those movements, but it seems far from revealing a physical cause for them. As projected against the celestial sphere the stars forming the two opposite streams appear intermingled, some obeying one tendency and some the other. As Professor Dyson has said, the hypothesis of this double movement is of a revolutionary character, and calls for further investigation. Indeed, it seems at first glance not less surprising than would be the observation that in a snowstorm the flakes over our heads were divided into two parties and driving across each other's course in nearly opposite directions, as if urged by interpenetrating winds.

But whatever explanation may eventually be found for the motions of the stars, the knowledge of the existence of those motions must always afford a new charm to the contemplative observer of the heavens, for they impart a sense of life to the starry system that would otherwise be lacking. A stagnant universe, with every star fixed immovably in its place, would not content the imagination or satisfy our longing for ceaseless activity. The majestic grandeur of the evolutions of the celestial hosts, the inconceivable vastness of the fields of space in which they are executed, the countless numbers, the immeasurable distances, the involved convolutions, the flocking and the scattering, the interpenetrating marches and counter-marches, the strange community of impulsion affecting stars that are wide apart in space and causing them to traverse the general movement about them like aides and dispatch-bearers on a battlefield — all these arouse an intensity of interest which is heightened by the mystery behind them.

The Passing of the Constellations

*F*rom a historical and picturesque point of view, one of the most striking results of the motions of the stars described in the last chapter is their effect upon the forms of the constellations, which have been watched and admired by mankind from a period so early that the date of their invention is now unknown. The constellations are formed by chance combinations of conspicuous stars, like figures in a kaleidoscope, and if our lives were commensurate with the æons of cosmic existence we should perceive that the kaleidoscope of the heavens was ceaselessly turning and throwing the stars into new symmetries. Even if the stars stood fast, the motion of the solar system would gradually alter the configurations, as the elements of a landscape dissolve and recombine in fresh groupings with the traveler's progress amid them. But with the stars themselves all in motion at various speeds and in many directions, the changes occur more rapidly. Of course, "rapid" is here understood in a relative sense; the wheel of human history to an eye accustomed to the majestic progression of the universe would appear to revolve with the velocity of a whirling dynamo. Only the deliberation of geological movements can be contrasted with the evolution and devolution of the constellations.

And yet this secular fluctuation of the constellation figures is not without keen interest for the meditative observer. It is another reminder of the swift mutability of terrestrial affairs.

To the passing glance, which is all that we can bestow upon these figures, they appear so immutable that they have been called into service to form the most lasting records of ancient thought and imagination that we possess. In the forms of the constellations, the most beautiful, and, in imaginative quality, the finest, mythology that the world has ever known has been perpetuated. Yet, in a broad sense, this scroll of human thought imprinted on the heavens is as evanescent as the summer clouds. Although more enduring than parchment, tombs, pyramids, and temples, it is as far as they from truly eternizing the memory of what man has fancied and done.

Before studying the effects that the motions of the stars have had and will have upon the constellations, it is worth while to consider a little further the importance of the stellar pictures as archives of history. To emphasize the importance of these effects it is only necessary to recall that the constellations register the oldest traditions of our race. In the history of primeval religions they are the most valuable of documents. Leaving out of account for the moment the more familiar mythology of the Greeks, based on something older yet, we may refer for illustration to that of the mysterious Maya race of America. At Izamal, in Yucatan, says Mr. Stansbury Hagar, is a group of ruins perched, after the Mexican and Central-American plan, on the summits of pyramidal mounds which mark the site of an ancient theogonic center of the Mayas. Here the temples all evidently refer to a cult based upon the constellations as symbols. The figures and the names, of course, were not the same as those that we have derived from our Aryan ancestors, but the star groups were the same or nearly so. For instance, the loftiest of the temples at Izamal was connected with the sign of the constellation known to us as Cancer, marking the place of the sun at the summer solstice, at which period the sun was supposed to descend at noon like a great bird of fire and consume the offerings left upon the altar. Our Scorpio was known to the Mayas as a sign of the "Death God." Our Libra, the "Balance," with which the idea of a divine weighing out of justice has always been connected, seems to be identical with the Mayan constellation Teoyaotlatohua, with which was associated a temple where dwelt the priests whose special business it was to administer justice and to foretell the future by means of information obtained from the spirits of the dead. Orion, the "Hunter" of our celestial mythology, was among the Mayas a "Warrior," while Sagittarius and others of our constellations were known to them (under different names, of course), and all were endowed with a religious symbolism. And the same star figures, having the same significance, were familiar to the Peruvians, as shown by the temples at Cuzco. Thus the imagination of ancient America sought in the constellations symbols of the unchanging gods.

But, in fact, there is no nation and no people that has not recognized the constellations, and at one period or another in its history employed them in some symbolic or representative capacity. As handled by the Greeks from prehistoric times, the constellation myths became the very soul of poetry. The imagination of that wonderful race idealized the principal star groups so effectively that the figures and traditions thus attached to them have, for civilized mankind, displaced all others, just as Greek art in its highest forms stands without parallel and eclipses every rival. The Romans translated no heroes and heroines of the mythical period of their history to the sky, and the deified Cæsars never entered that lofty company, but the heavens are filled with the early myths of the Greeks. Herakles nightly resumes his mighty labors in the stars; Zeus, in the form of the white "Bull," Taurus, bears the fair Europa on his back through the celestial waves; Andromeda stretches forth her shackled arms in the star-gemmed ether, beseeching aid; and Perseus, in a blaze of diamond armor, revives his heroic deeds amid sparkling clouds of stellar dust. There, too, sits Queen Cassiopeia in her dazzling chair, while the Great King, Cepheus, towers gigantic over the pole. Professor Young has significantly remarked that a great number of the constellations are connected in some way or other with the Argonautic Expedition — that strangely fascinating legend of earliest Greek story which has never lost its charm for mankind. In view of all this, we may well congratulate ourselves that the constellations will outlast our time and the time of countless generations to follow us; and yet they are very far from being eternal.

Let us now study some of the effects of the stellar motions upon them.

We begin with the familiar figure of the "Great Dipper." He who has not drunk inspiration from its celestial bowl is not yet admitted to the circle of Olympus. This figure is made up of seven conspicuous stars in the constellation Ursa Major, the "Greater Bear." The handle of the "Dipper" corresponds to the tail of the imaginary "Bear," and the bowl lies upon his flank. In fact, the figure of a dipper is so evident and that of a bear so unevident, that to most persons the "Great Dipper" is the only part of the constellation that is recognizable. Of the seven stars mentioned, six are of nearly equal brightness, ranking as of the second magnitude, while the seventh is of only the third magnitude. The difference is very striking, since every increase of one magnitude involves an increase of two-and-a-half times in brightness. There appears to be little doubt that the faint star, which is situated at the junction of the bowl and the handle, is a variable of long period, since three hundred years ago it was as bright as its companions. But however that may be, its relative faintness at the present time interferes but little with the perfection of the "Dipper's" figure. In order the more readily to understand the changes which are taking place, it will be well to mention both the names and the Greek letters which are attached to the seven stars. Beginning at the star in the upper outer edge of the rim of the bowl and running in regular order round the bottom and then out to the end of the handle, the names and letters are as follows: Dubhe (α), Merak (β), Phaed (χ), Megrez (δ), Alioth (ϵ), Mizar (ϕ), and Benetnasch (γ). Megrez is the faint star already mentioned at the junction of the bowl and handle, and Mizar, in the middle of the handle, has a close, naked-eye companion which is named Alcor. The Arabs called this singular pair of stars "The Horse and Rider." Merak and Duhbe are called "The Pointers," because an imaginary line drawn northward through them indicates the Pole Star.

Now it has been found that five of these stars — *viz.*, Merak, Phaed, Megrez, Alioth, and Mizar (with its comrade) — are moving with practically the same speed in an easterly direction, while the other two, Dubhe and Benetnasch, are simultaneously moving westward, the motions of Benetnasch being apparently more rapid. The consequence of these opposed motions is, of course, that the figure of the "Dipper" cannot always have existed and will not continue to exist. In the accompanying diagrams it has been thought interesting to show the relative positions of these seven stars, as seen from the point which the earth now occupies, both in the past and in the future. Arrows attached to the stars in the figure representing the present appearance of the "Dipper" indicate the directions of the motions and the distances over which they will carry the stars in a period of about five hundred centuries. The time, no doubt, seems long, but remember the vast stretch of ages through which the earth has passed, and then reflect that no reason is apparent why our globe should not continue to be a scene of animation for ten thousand centuries yet to come. The fact that the little star Alcor placed so close to Mizar should accompany the latter in its flight is not surprising, but that two of the principal stars of the group should be found moving in a direction directly opposed to that pursued by the other five is surprising in the highest degree; and it recalls the strange theory of a double drift affecting all the stars, to which attention was called in the preceding chapter. It would appear that Benetnasch and Dubhe belong to one "current," and Merak, Phaed, Megrez, Alioth, and Mizar to the other. As far as is known, the motion of the seven stars are not shared by the smaller stars scattered about them, but on the theory of currents there should be such a community of motion, and further investigation may reveal it.

From the "Great Dipper" we turn to a constellation hardly less conspicuous and situated at an equal distance from the pole on the other side — Cassiopeia. This famous star-group commemorating the romantic Queen of Ethiopia whose vain boasting of her beauty was punished by the exposure of her daughter Andromeda to the "Sea Monster," is well-marked by five stars which form an irregular letter "W" with its open side toward the pole. Three of these stars are usually ranked as of the second magnitude, and two of the third; but to ordinary observation they appear of nearly equal brightness, and present a very striking picture. They mark out the chair and a part of the figure of the beautiful queen. Beginning at the right-hand, or western, end of the "W," their Greek letter designations are: Beta (β), Alpha (α), Gamma (χ), Delta (d), and Epsilon (ε). Four of them, Beta, Alpha, Delta, and Epsilon are traveling eastwardly at various speeds, while the fifth, Gamma, moves in a westerly direction. The motion of Beta is more rapid than that of any of the others. It should be said, however, that no little uncertainty attaches to the estimates of the rate of motion of stars which are not going very rapidly, and different observers often vary considerably in their results.

In the beautiful "Northern Crown," one of the most perfect and charming of all the figures to be found in the stars, the alternate combining and scattering effects of the stellar motions are shown by comparing the appearance which the constellation must have had five hundred centuries ago with that which it has at present and that which it will have in the future. The seven principle stars of the asterism, forming a surprisingly perfect coronet, have movements in three directions at right angles to one another. That in these circumstances they should ever have arrived at positions giving them so striking an appearance of definite association is certainly surprising; from its aspect one would have expected to find a community of movement governing the brilliants of the "Crown," but instead of that we find evidence that they will inevitably drift apart and the beautiful figure will dissolve.

A similar fate awaits such asterisms as the "Northern Cross" in Cygnus; the "Crow" (Corvus), which stands on the back of the great "Sea Serpent," Hydra, and pecks at his scales; "Job's Coffin" (Delphinus); the "Great Square of Pegasus"; the "Twins" (Gemini); the beautiful "Sickle" in Leo; and the exquisite group of the Hyades in Taurus. In the case of the Hyades, two controlling movements are manifest: one, affecting five of the stars which form the well-known figure of a letter "V," is directed northerly; the other, which controls the direction of two stars, has an easterly trend. The chief star of the group, Aldebaran, one of the finest of all stars both for its brilliance and its color, is the most affected by the easterly motion. In time it will drift entirely out of connection with its present neighbors. Although the Hyades do not form so compact a group as the Pleiades in the same constellation, yet their appearance of relationship is sufficient to awaken a feeling of surprise over the fact that, as with the stars of the "Dipper," their association is only temporary or apparent.

The great figure of Orion appears to be more lasting, not because its stars are physically connected, but because of their great distance, which renders their movements too deliberate to be exactly ascertained. Two of the greatest of its stars, Betelgeuse and Rigel, possess, as far as has been ascertained, no perceptible motion across the line of sight, but there is a little movement perceptible in the "Belt." At the present time this consists of an almost perfect straight line, a row of second-magnitude stars about equally spaced and of the most striking beauty. In the course of time, however, the two right-hand stars, Mintaka and Alnilam (how fine are these Arabic star names!) will approach each other and form a naked-eye double, but the third, Alnita, will drift away eastward, so that the "Belt" will no longer exist.

For one more example, let us go to the southern hemisphere, whose most celebrated constellation, the "Southern Cross," has found a place in all modern literatures, although it has no claim to consideration on account of association with ancient legends. This most attractive asterism, which has never ceased to fascinate the imagination of Christendom since it was first devoutly described by the early explorers of the South, is but a passing collocation of brilliant stars. Yet even in its transfigurations it has been for hundreds of centuries, and will continue to be for hundreds of centuries to come, a most striking object in the sky. Our figures show its appearance in three successive phases: first, as it was fifty thousand years ago (viewed from the earth's present location); second, as it is in our day; and, third, as it will be an equal time in the future. The nearness of these bright stars to one another — the length of the longer beam of the "Cross" is only six degrees — makes this group very noticeable, whatever the arrangement of its components may be. The largest star, at the base of the "Cross," is of the first magnitude, two of the others are of the second magnitude, and the fourth is of the third. Other stars, not represented in the figures, increase the effect of a celestial blazonry, although they do not help the resemblance to a cross.

But since the motion of the solar system itself will, in the course of so long a period as fifty thousand years, produce a great change in the perspective of the heavens as seen from the earth, by carrying us nearly nineteen trillion miles from our present place, why, it may be asked, seek to represent future appearances of the constellations which we could not hope to see, even if we could survive so long? The answer is: Because these things aid the mind to form a picture of the effects of the mobility of the starry universe. Only by showing the changes from some definite point of view can we arrive at a due comprehension of them. The constellations are more or less familiar to everybody, so that impending changes of their forms must at once strike the eye and the imagination, and make clearer the significance of the movements of the stars. If the future history of mankind is to resemble its past and if our race is destined to survive yet a million years, then our remote descendents will see a "new heavens" if not a "new earth," and will have to invent novel constellations to perpetuate their legends and mythologies.

If our knowledge of the relative distances of the stars were more complete, it would be an interesting exercise in celestial geometry to project the constellations probably visible to the inhabitants of worlds revolving around some of the other suns of space. Our sun is too insignificant for us to think that he can make a conspicuous appearance among them, except, perhaps, in a few cases. As seen, for instance, from the nearest known star, Alpha Centauri, the sun would appear of the average first magnitude, and consequently from that standpoint he might be the gem of some little constellation which had no Sirius, or Arcturus, or Vega to eclipse him with its superior splendor. But from the distance of the vast majority of the stars the sun would probably be invisible to the naked eye, and as seen from nearer systems could only rank as a fifth or sixth magnitude star, unnoticed and unknown except by the star-charting astronomer.

Conflagrations in the Heavens

Suppose it were possible for the world to take fire and burn up — as some pessimists think that it will do when the Divine wrath shall have sufficiently accumulated against it — nobody out of our own little corner of space would ever be aware of the catastrophe! With all their telescopes, the astronomers living in the golden light of Arcturus or the diamond blaze of Canopus would be unable to detect the least glimmer of the conflagration that had destroyed the seat of Adam and his descendents, just as now they are totally ignorant of its existence.

But at least fifteen times in the course of recorded history men looking out from the earth have beheld in the remote depths of space great outbursts of fiery light, some of them more splendidly luminous than anything else in the firmament except the sun! If *they* were conflagrations, how many million worlds like ours were required to feed their blaze?

It is probable that "temporary" or "new" stars, as these wonderful apparitions are called, really are conflagrations; not in the sense of a bonfire or a burning house or city, but in that of a sudden eruption of inconceivable heat and light, such as would result from the stripping off the shell of an encrusted sun or the crashing together of two mighty orbs flying through space with a hundred times the velocity of the swiftest cannon-shot.

Temporary stars are the rarest and most erratic of astronomical phenomena. The earliest records relating to them are not very clear, and we cannot in every instance be certain that it was one of these appearances that the ignorant and super-

stitious old chroniclers are trying to describe. The first temporary star that we are absolutely sure of appeared in 1572,
and is known as "Tycho's Star," because the celebrated Danish
astronomer (whose remains, with his gold-and-silver artificial
nose — made necessary by a duel — still intact, were disinterred
and reburied in 1901) was the first to perceive it in the sky,
and the most assiduous and successful in his studies of it. As
the first fully accredited representative of its class, this new
star made its entry upon the scene with becoming *éclat.* It is
characteristic of these phenomena that they burst into view
with amazing suddenness, and, of course, entirely unexpect-
edly. Tycho's star appeared in the constellation Cassiopeia,
near a now well-known and much-watched little star named
Kappa, on the evening of November 11, 1572. The story has
often been repeated, but it never loses interest, how Tycho,
going home that evening, saw people in the street pointing
and staring at the sky directly over their heads, and following
the direction of their hands and eyes he was astonished to
see, near the zenith, an unknown star of surpassing brilliance.
It outshone the planet Jupiter, and was therefore far brighter
than the first magnitude. There was not another star in the
heavens that could be compared with it in splendor. Tycho
was not in all respects free from the superstitions of his time
— and who is? — but he had the true scientific instinct, and
immediately he began to study the stranger, and to record
with the greatest care every change in its aspect. First he
determined as well as he could with the imperfect instruments
of his day, many of which he himself had invented, the precise
location of the phenomena in the sky. Then he followed the
changes that it underwent. At first it brightened until its light
equaled or exceeded that of the planet Venus at her brightest,
a statement which will be appreciated at its full value by
anyone who has ever watched Venus when she plays her
dazzling rôle of "Evening Star," flaring like an arc light in
the sunset sky. It even became so brilliant as to be visible in
full daylight, since, its position being circumpolar, it never
set in the latitude of Northern Europe. Finally it began to
fade, turning red as it did so, and in March, 1574, it disappeared from Tycho's searching gaze, and has never been seen
again from that day to this. None of the astronomers of the

time could make anything of it. They had not yet as many bases of speculation as we possess today.

Tycho's star has achieved a romantic reputation by being fancifully identified with the "Star of Bethlehem," said to have led the wondering Magi from their eastern deserts to the cradle-manger of the Savior in Palestine. Many attempts have been made to connect this traditional "star" with some known phenomenon of the heavens, and none seems more idle than this. Yet it persistently survives, and no astronomer is free from eager questions about it addressed by people whose imagination has been excited by the legend. It is only necessary to say that the supposition of a connection between the phenomenon of the Magi and Tycho's star is without any scientific foundation. It was originally based on an unwarranted assumption that the star of Tycho was a variable of long period, appearing once every three hundred and fifteen years, or thereabout. If that were true there would have been an apparition somewhere near the traditional date of the birth of Christ, a date which is itself uncertain. But even the data on which the assumption was based are inconsistent with the theory. Certain monkish records speak of something wonderful appearing in the sky in the years 1264 and 945, and these were taken to have been outbursts of Tycho's star. Investigation shows that the records more probably refer to comets, but even if the objects seen were temporary stars, their dates do not suit the hypothesis; from 945 to 1264 there is a gap of 319 years, and from 1264 to 1572 one of only 308 years; moreover 337 years have now (1909) elapsed since Tycho saw the last glimmer of his star. Upon a variability so irregular and uncertain as that, even if we felt sure that it existed, no conclusion could be found concerning an apparition occurring 2000 years ago.

In the year 1600 (the year in which Giordano Bruno was burned at the stake for teaching that there is more than one physical world), a temporary star of the third magnitude broke out in the constellation Cygnus, and curiously enough, considering the rarity of such phenomena, only four years later another surprisingly brilliant one appeared in the constellation Ophiuchus. This is often called "Kepler's star," because the great German astronomer devoted to it the same attention that Tycho had given to the earlier phenomenon. It, too, like Tycho's, was at first the brightest object in the stellar heavens, although it seems never to have quite equaled its famous predecessor in splendor. It disappeared after a year, also turning of a red color as it became more faint. We shall see the significance of this as we go on. Some of Kepler's contemporaries suggested that the outburst of this star was due to a meeting of atoms in space, and idea bearing a striking resemblance to the modern theory of "astronomical colli-sions."

In 1670, 1848, and 1860 temporary stars made their ap-pearance, but none of them was of great brilliance. In 1866 one of the second magnitude broke forth in the "Northern Crown" and awoke much interest, because by that time the spectroscope had begun to be employed in studying the composition of the stars, and Huggins demonstrated that the new star consisted largely of incandescent hydrogen. But this star, apparently unlike the others mentioned, was not abso-lutely new. Before its outburst it had shown as a star of the ninth magnitude (entirely invisible, of course, to the naked eye), and after about six weeks it faded to its original condi-tion in which it has ever since remained. In 1876 a temporary star appeared in the constellation Cygnus, and attained at one time the brightness of the second magnitude. Its spec-trum and its behavior resembled those of its immediate predecessor. In 1885, astronomers were surprised to see a sixth-magnitude star glimmering in the midst of the hazy cloud of the great Andromeda Nebula. It soon absolutely disappeared. Its spectrum was remarkable for being "continu-ous," like that of the nebula itself. A continuous spectrum is supposed to represent a body, or a mass, which is either solid or liquid, or composed of gas under great pressure. In January, 1892, a new star was suddenly seen in the constellation Auriga. It never rose much above the fourth magnitude, but it showed a peculiar spectrum containing both bright and dark lines of hydrogen.

But a bewildering surprise was now in store; the world was to behold at the opening of the twentieth century such a celestial spectacle as had not been on view since the times of Tycho and Kepler. Before daylight on the morning of February 22, 1901, the Rev. Doctor Anderson, of Edinburgh, an amateur astronomer, who had also been the first to see the new star in Auriga, beheld a strange object in the constellation Perseus not far from the celebrated variable star Algol. He recognized its character at once, and immediately telegraphed the news, which awoke the startled attention of astronomers all over the world. When first seen the new star was no brighter than Algol (less than the second magnitude), but within twenty-four hours it was ablaze, outshining even the brilliant Capella, and far surpassing the first magnitude. At the spot in the sky where it appeared nothing whatever was visible on the night before its coming. This is known with certainty because a photograph had been made of that very region on February 21, and this photograph showed every-thing down to the twelfth magnitude, but not a trace of the stranger which burst into view between the 21st and the 22nd like the explosion of a rocket.

Upon one who knew the stars the apparition of this in-truder in a well-known constellation had the effect of a sudden invasion. The new star was not far west of the zenith in the early evening, and in that position showed to the best advantage. To see Capella, the hitherto unchallenged ruler of that quarter of the sky, abased by comparison with this stranger of alien aspect, for there was always an unfamiliar look about the "nova," was decidedly disconcerting. It seemed to portend the beginning of a revolution in the heavens. One could understand what the effect of such an apparition must have been in the superstitious times of Tycho. The star of Tycho had burst forth on the northern border of the Milky Way; this one was on its southern border, some forty-five degrees farther east.

Astronomers were well-prepared this time for the scientific study of the new star, both astronomical photography and spectroscopy having been perfected, and the results of their investigations were calculated to increase the wonder with which the phenomenon was regarded. The star remained at its brightest only a few days; then, like a veritable conflagra-tion, it began to languish; and, like the reflection of a dying fire, as it sank it began to glow with the red color of embers. But its changes were spasmodic; once about every three days it flared up only to die away again. During these fluctuations its light varied alternately in the ratio of one to six. Finally it took a permanent downward course, and after a few months the naked eye could no longer perceive it; but it remained visible with telescopes, gradually fading until it had sunk to the ninth magnitude. Then another astonishing change happened: in August photographs taken at the Yerkes Observatory and at Heidelberg showed that the "nova" was *surrounded by a spiral nebula!* The nebula had not been there before, and no one could doubt that it represented a phase of the same catastrophe that had produced the outburst of the new star. At one time the star seemed virtually to have disappeared, as if all its substance had been expanded into the nebulous cloud, but always there remained a stellar nucleus about which the misty spiral spread wider and ever wider, like a wave expanding around a center of disturbance. The nebula too showed a variability of brightness, and four condensations which formed in it seemed to have a motion of revolution about the star. As time went on the nebula continued to expand at a rate which was computed to be not less than twenty thousand miles per second! And now the star itself, showing indications of having turned into a nebula, behaved in a most erratic manner, giving rise to the suspicion that it was about to burst out again. But this did not occur, and at length it sunk into a state of lethargy from which it has to the present time not recovered. But the nebulous spiral has disappeared, and the entire phenomena as it now (1909) exists consists of a faint nebulous star of less than the ninth magnitude.

The wonderful transformations just described had been forecast in advance of the discovery of the nebulous spiral encircling the star by the spectroscopic study of the latter. At first there was no suggestion of a nebular constitution, but within a month or two characteristic nebular lines began to appear, and in less than six months the whole spectrum had been transformed to the nebular type. In the meantime the shifting of the spectral lines indicated a complication of rapid motions in several directions simultaneously. These motions were estimated to amount to from one hundred to five hundred miles per second.

The human mind is so constituted that it feels forced to seek an explanation of so marvelous a phenomenon as this, even in the absence of the data needed for a sound conclusion. The most natural hypothesis, perhaps, is that of a collision. Such a catastrophe could certainly happen. It has been shown, for instance, that in infinity of time the earth is sure to be hit by a comet; in the same way it may be asserted that, if no time limit is fixed, the sun is certain to run against some obstacle in space, either another star, or a dense meteor swarm, or one of the dark bodies which there is every reason to believe abound around us. The consequences of such a collision are easy to foretell, provided that we know the masses and the velocities of the colliding bodies. In a preceding chapter we have discussed the motions of the sun and stars, and have seen that they are so swift that an encounter between any two of them could not but be disastrous. But this is not all; for as soon as two stars approached within a few million miles their speed would be enormously increased by their reciprocal attractions and, if their motion was directed radially with respect to their centers, they would come together with a crash that would reduce them both to nebulous clouds. It is true that the chances of such a "head-on" collision are relatively very small; two stars approaching each other would most probably fall into closed orbits around their common center of gravity. If there were a collision it would most likely be a grazing one instead of a direct front-to-front encounter. But even a close approach, without any actual collision, would probably prove disastrous, owing to the tidal influence of each of the bodies on the other. Suns, in consequence of their enormous masses and dimensions and the peculiarities of their constitution, are exceedingly dangerous to one another at close quarters. Propinquity awakes in them a mutually destructive tendency. Consisting of matter in the gaseous, or perhaps, in some cases, liquid, state, their tidal pull upon each other if brought close together might burst them asunder, and the photospheric envelope being destroyed the internal incandescent mass would gush out, bringing fiery death to any planets that were revolving near. Without regard to the resulting disturbance of the earth's orbit, the close approach of a great star to the sun would be in the highest degree perilous to us. But this is a danger which may properly be regarded as indefinitely remote, since, at our present location in space, we are certainly far from every star except the sun, and we may feel confident that no great invisible body is near, for if there were one we should be aware of its presence from the effects of its attraction. As to dark nebulæ which may possibly lie in the track that the solar system is pursuing at the rate of 375,000,000 miles per year, that is another question — and they, too, could be dangerous!

This brings us directly back to "Nova Persei," for among the many suggestions offered to explain its outburst, as well as those of other temporary stars, one of the most fruitful is that of a collision between a star and a vast invisible nebula. Professor Seeliger, of Munich, first proposed this theory, but it afterward underwent some modifications from others. Stated in a general form, the idea is that a huge dark body, perhaps an extinguished sun, encountered in its progress through space a widespread flock of small meteors forming a dark nebula. As it plunged into the swarm the friction of the innumerable collisions with the meteors heated its surface to incandescence, and being of vast size it then became visible to us as a new star. Meanwhile the motion of the body through the nebula, and its rotation upon itself, set up a gyration in the blazing atmosphere formed around it by the vaporized meteors; and as this atmosphere spread wider, under the laws of gyratory motion a rotation in the opposite direction began in the inflamed meteoric cloud outside the central part of the vortex. Thus the spectral lines were caused to show motion in opposite directions, a part of the incandescent mass approaching the earth simultaneously with the retreat of another part. So the curious spectroscopic observations before mentioned were explained. This theory might also account for the appearance of the nebulous spiral first seen some six months after the original outburst. The sequent changes in the spectrum of the "nova" are accounted for by this theory on the assumption, reasonable enough in itself, that at first the invading body would be enveloped in a vaporized atmosphere of relatively slight depth, producing by its absorption the fine dark lines first observed; but that as time went on and the incessant collisions continued, the blazing atmosphere would become very deep and extensive, whereupon the appearance of the spectral lines would change, and bright lines due to the light of the incandescent meteors surrounding the nucleus at a great distance would take the place of the original dark ones. The vortex of meteors once formed would protect the flying body within from further immediate collisions, the latter now occurring mainly among the meteors themselves, and then the central blaze would die down, and the original splendor of the phenomenon would fade.

But the theories about Nova Persei have been almost as numerous as the astronomers who have speculated about it. One of the most startling of them assumed that the outburst was caused by the running amuck of a dark star which had encountered another star surrounded with planets, the re-newed outbreaks of light after the principal one had faded being due to the successive running down of the unfortunate planets! Yet another hypothesis is based on what we have already said of the tidal influence that two close approaching suns would have upon each other. Supposing two such bodies which had become encrusted, but remained incandescent and fluid within, to approach within almost striking distance; they would whirl each other about their common center of gravity, and at the same time their shells would burst under the tidal strain, and their glowing nuclei being disclosed would produce a great outburst of light. Applying this theory to a "nova," like that of 1866 in the "Northern Crown," which had been visible as a small star before the outbreak, and which afterward resumed its former aspect, we should have to as-sume that a yet shining sun had been approached by a dark body whose attraction temporarily burst open its photo-sphere. It might be supposed that in this case the dark body was too far advanced in cooling to suffer the same fate from the tidal pull of its victim. But a close approach of that kind would be expected to result in the formation of a binary system, with orbits of great eccentricity, perhaps, and after the lapse of a certain time the outburst should be renewed by another approximation of the two bodies. A temporary star of that kind would rather be ranked as a variable.

The celebrated French astronomer, Janssen, had a different theory of Nova Persei, and of temporary stars in general. According to his idea, such phenomena might be the result of chemical changes taking place in a sun without interference by, or collision with, another body. Janssen was engaged for many years in trying to discover evidence of the existence of oxygen in the sun, and he constructed his observatory on the summit of Mount Blanc specially to pursue that research. He believed that oxygen must surely exist in the sun since we find so many other familiar elements included in the constitution of the solar globe, and as he was unable to discover satisfactory evidence of its presence he assumed that it existed in a form unknown on the earth. If it were normally in the sun's chromosphere, or coronal atmosphere, he said, it would combine with the hydrogen which we know is there and form an obscuring envelope of water vapor. It exists, then, in a special state, uncombined with hydrogen; but let the temperature of the sun sink to a critical point and the oxygen will assume its normal properties and combine with the hydrogen, producing a mighty outburst of light and heat. This, Janssen thought, might explain the phenomena of the temporary stars. It would also, he suggested, account for their brief career, because the combination of the elements would be quickly accomplished, and then the resulting water vapor would form an atmosphere cutting off the radiation from the star within.

This theory may be said to have a livelier human interest than some of the others, since, according to it, the sun may carry in its very constitution a menace to mankind; one does not like to think of it being suddenly transformed into a gigantic laboratory for the explosive combination of oxygen and hydrogen! But while Janssen's theory might do for some temporary stars, it is inadequate to explain all the phenomena of Nova Persei, and particularly the appearance of the great spiral nebula that seemed to exhale from the heart of the star. Upon the whole, the theory of an encounter between a star and a dark nebula seems best to fit the observations. By that hypothesis the expanding billow of light surrounding the core of the conflagration is very well accounted for, and the spectroscopic peculiarities are also explained.

Dr. Gustov Le Bon offers a yet more alarming theory, suggesting that temporary stars are the result of *atomic explosion;* but we shall touch upon this more fully in Chapter 14.

Twice in the course of this discussion we have called attention to the change of color invariably undergone by temporary stars in the later stages of their career. This was conspicuous with Nova Persei which glowed more and more redly as it faded, until the nebulous light began to overpower that of the stellar nucleus. Nothing could be more suggestive of the dying out of a great fire. Moreover, change of color from white to red is characteristic of all variable stars of long period, such as "Mira" in Cetus. It is also characteristic of stars believed to be in the later stages of evolution, and consequently approaching extinction, like Antares and Betelgeuse, and still more notably certain small stars which "gleam like rubies in the field of the telescope." These last appear to be suns in the closing period of existence as self-luminous bodies. Between the white stars, such as Sirius and Rigel, and the red stars, such as Aldebaran and Alpha Herculis, there is a progressive series of colors from golden yellow through orange to deep red. The change is believed to be due to the increase of absorbing vapors in the stellar atmosphere as the body cools down. In the case of ordinary stars these changes no doubt occupy many millions of years, which represent the average duration of solar life; but the temporary stars run through similar changes in a few months: they resemble ephemeral insects — born in the morning and doomed to perish with the going down of the sun.

Explosive and Whirling Nebulæ

One of the most surprising triumphs of celestial photography was Professor Keeler's discovery, in 1899, that the great majority of the nebulæ have a distinctly spiral form. This form, previously known in Lord Rosse's great "Whirlpool Nebula," had been supposed to be exceptional; now the photographs, far excelling telescopic views in the revelation of nebular forms, showed the spiral to be the typical shape. Indeed, it is a question whether all nebulæ are not to some extent spiral. The extreme importance of this discovery is shown in the effect that it has had upon hitherto prevailing views of solar and planetary evolution. For more than three-quarters of a century Laplace's celebrated hypothesis of the manner of origin of the solar system from a rotating and contracting nebula surrounding the sun had guided speculation on that subject, and had been tentatively extended to cover the evolution of systems in general. The apparent forms of some of the nebulæ which the telescope had revealed were regarded, and by some are still regarded, as giving visual evidence in favor of this theory. There is a "ring nebula" in Lyra with a central star, and a "planetary nebula" in Gemini bearing no little resemblance to the planet Saturn with its rings, both of which appear to be practical realizations of Laplace's idea, and the elliptical rings surrounding the central condensation of the Andromeda Nebula may be cited for the same kind of proof.

But since Keeler's discovery there has been a decided turning away of speculation another way. The form of the spiral nebulæ seems to be entirely inconsistent with the theory of an originally globular or disk-shaped nebula condensing around a sun and throwing or leaving off rings, to be subsequently shaped into planets. Some astronomers, indeed, now reject Laplace's hypothesis *in toto,* preferring to think that even our solar system originated from a spiral nebula. Since the spiral type prevails among the existing nebulæ, we must make any mechanical theory of the development of stars and planetary systems from them accord with the requirements which that form imposes. A glance at the extraordinary variations upon the spiral which Professor Keeler's photographs reveal is sufficient to convince one of the difficulty of the task of basing a general theory upon them. In truth, it is much easier to criticize Laplace's hypothesis than to invent a satisfactory substitute for it. If the spiral nebulæ seem to oppose it there are other nebulæ which appear to support it, and it may be that no one fixed theory can account for all the forms of stellar evolution in the universe. Our particular planetary system may have originated very much as the great French mathematician supposed, while others have undergone, or are now undergoing, a different process of development. There is always a too strong tendency to regard an important new discovery and the theories and speculations based upon it as revolutionizing knowledge, and displacing or overthrowing everything that went before. Upon the plea that "Laplace only made a guess" more recent guesses have been driven to extremes and treated by injudicious exponents as "the solid facts at last."

Before considering more recent theories than Laplace's, let us see what the nature of the photographic revelations is. The vast celestial maelstrom discovered by Lord Rosse in the "Hunting Dogs" may be taken as the leading type of the spiral nebulæ, although there are less conspicuous objects of the kind which, perhaps, better illustrate some of their peculiarities. Lord Rosse's nebula appears far more wonderful in the photographs than in his drawings made with the aid of his giant reflecting telescope at Parsonstown, for the photographic plate records details that no telescope is capable of showing. Suppose we look at the photograph of this object as any person of common sense would look at any great and strange natural phenomenon. What is the first thing that strikes the mind? It is certainly the appearance of violent whirling motion. One would say that the whole glowing mass had been spun about with tremendous velocity, or that it had been set rotating so rapidly that it had become the victim of "centrifugal force," one huge fragment having broken loose and started to gyrate off into space. Closer inspection shows that in addition to the principal focus there are various smaller condensations scattered through the mass. These are conspicuous in the spirals. Some of them are stellar points, and but for the significance of their location we might suppose them to be stars which happen to lie in a line between us and the nebula. But when we observe how many of them follow most faithfully the curves of the spirals we cannot but conclude that they form an essential part of the phenomenon; it is not possible to believe that their presence in such situations is merely fortuitous. One of the outer spirals has at least a dozen of these starlike points strung upon it; some of them sharp, small, and distinct, others more blurred and nebulous, suggesting different stages of condensation. Even the part which seems to have been flung loose from the main mass has, in addition to its central condensation, at least one stellar point gleaming in the half-vanished spire attached to it. Some of the more distant stars scattered around the "whirl-pool" look as if they too had been shot out of the mighty vortex, afterward condensing into unmistakable solar bodies. There are at least two curved rows of minute stars a little beyond the periphery of the luminous whirl which clearly follow lines concentric with those of the nebulous spirals. Such facts are simply dumbfounding for anyone who will bestow sufficient thought upon them, for these are *suns*, though they may be small ones; and what a birth is that for a sun!

Look now again at the glowing spirals. We observe that hardly have they left the central mass before they begin to coagulate. In some places they have a "ropy" aspect; or they are like peascods filled with growing seeds, which eventually will become stars. The great focus itself shows a similar tendency, especially around its circumference. The sense that it imparts of a tremendous shattering force at work is overwhelming. There is probably more matter in that whirling and bursting nebula than would suffice to make a hundred solar systems! It must be confessed at once that there is no confirmation of the Laplacean hypothesis here; but what hypothesis will fit the facts? There is one which it has been claimed does so, but we shall come to that later. In the meanwhile, as a preparation, fix in the memory the appearance of that second spiral mass spinning beside its master which seems to have spurned it away.

For a second example of the spiral nebulæ look at the one in the constellation Triangulum. *God, how hath the imagination of puny man failed to comprehend Thee!* Here is creation through destruction with a vengeance! The spiral form of the nebula is unmistakable, but it is half obliterated amid the turmoil of flying masses hurled away on all sides with tornadic fury. The focus itself is splitting asunder under the intolerable strain, and in a little while, as time is reckoned in the Cosmos, it will be gyrating into stars. And then look at the cyclonic rain of already finished stars whirling round the outskirts of the storm. Observe how scores of them are yet involved in the fading streams of the nebulous spirals; see how they have been thrown into vast loops and curves, of a beauty that half redeems the terror of the spectacle enclosed within their lines — like iridescent cirri hovering about the edges of a hurricane. And so again are suns born!

Let us turn to the exquisite spiral in Ursa Major; how different its aspect from that of the other! One would say that if the terrific coil in Triangulum has all but destroyed itself in its fury, this one on the contrary has just begun its self-demolition. As one gazes one seems to see in it the smooth, swift, accelerating motion that precedes catastrophe. The central part is still intact, dense, and uniform in texture. How graceful are the spirals that smoothly rise from its oval rim and, gemmed with little stars, wind off into the darkness until they have become as delicate as threads of gossamer! But at bottom the story told here is the same — creation by gyration!

Compare with the above the curious mass in Cetus. Here the plane of the whirling nebula nearly coincides with our line of sight and we see the object at a low angle. It is far advanced and torn to shreds, and if we could look at it perpendicularly to its plane it is evident that it would closely resemble the spectacle in Triangulum.

Then take the famous Andromeda Nebula (see Frontispiece), which is so vast that notwithstanding its immense distance even the naked eye perceives it as an enigmatical wisp in the sky. Its image on the sensitive plate is the masterpiece of astronomical photography; for wild, incomprehensible beauty there is nothing that can be compared with it. Here, if anywhere, we look upon the spectacle of creation in one of its earliest stages. The Andromeda Nebula is apparently less advanced toward transformation into stellar bodies than is that in Triangulum. The immense crowd of stars sprinkled over it and its neighborhood seem in the main to lie this side of the nebula, and consequently to have no connection with it. But incipient stars (in some places clusters of them) are seen in the nebulous rings, while one or two huge masses seem to give promise of transformation into stellar bodies of unusual magnitude. I say "rings" because although the loops encompassing the Andromeda Nebula have been called spirals by those who wish utterly to demolish Laplace's hypothesis, yet they are not manifestly such, as can be seen on comparing them with the undoubted spirals of the Lord Rosse Nebula. They look quite as much like circles or ellipses seen at an angle of, say, fifteen or twenty degrees to their plane. If they are truly elliptical they accord fairly well with Laplace's idea, except that the scale of magnitude is stupendous, and if the Andromeda Nebula is to become a solar system it will surpass ours in grandeur beyond all possibility of comparison.

There is one circumstance connected with the spiral nebulæ, and conspicuous in the Andromeda Nebula on account of its brightness, which makes the question of their origin still more puzzling; they all show continuous spectra, which, as we have before remarked, indicate that the mass from which the light comes is either solid or liquid, or a gas under heavy pressure. Thus nebulæ fall into two classes: the "white"

nebulæ, giving a continuous spectrum; and the "green" nebulæ whose spectra are distinctly gaseous. The Andromeda
Nebula is the great representative of the former class and the
Orion Nebula of the latter. The spectrum of the Andromeda
Nebula has been interpreted to mean that it consists not of
luminous gas, but of a flock of stars so distant that they are
separately indistinguishable even with powerful telescopes,
just as the component stars of the Milky Way are indistinguishable with the naked eye; and upon this has been based
the suggestion that what we see in Andromeda is an outer
universe whose stars form a series of elliptical garlands surrounding a central mass of amazing richness. But this idea is
unacceptable if for no other reason than that, as just said, all
the spiral nebulæ possess the same kind of spectrum, and
probably no one would be disposed to regard them all as outer
universes. As we shall see later, the peculiarity of the spectra
of the spiral nebulæ is appealed to in support of a modern
substitute for Laplace's hypothesis.

Finally, without having by any means exhausted the variety
exhibited by the spiral nebulæ, let us turn to the great representative of the other species, the Orion Nebula. In some
ways this is even more marvelous than the others. The early
drawings with the telescope failed to convey an adequate
conception either of its sublimity or of its complication of
structure. It exists in a nebulous region of space, since photographs show that nearly the whole constellation is inter-
woven with faintly luminous coils. To behold the entry of the
great nebula into the field even of a small telescope is a
startling experience which never loses its novelty. As shown
by the photographs, it is an inscrutable chaos of perfectly
amazing extent, where spiral bands, radiating streaks, dense
masses, and dark yawning gaps are strangely intermingled
without apparent order. In one place four conspicuous little
stars, better seen in a telescope than in the photograph on
account of the blurring produced by overexposure, are suggestively situated in the midst of a dark opening, and no
observer has ever felt any doubt that these stars have been
formed from the substance of the surrounding nebula. There
are many other stars scattered over its expanse which manifestly owe their origin to the same source. But compare the

general appearance of this nebula with the others that we have studied, and remark the difference. If the unmistakably spiral nebulæ resemble bursting fly-wheels or grindstones from whose perimeters torrents of sparks are flying, the Orion Nebula rather recalls the aspect of a cloud of smoke and fragments produced by the explosion of a shell. This idea is enforced by the look of the outer portion farthest from the bright half of the nebula, where sharply edged clouds with dark spaces behind seem to be billowing away as if driven by a wind blowing from the center.

Next let us consider what scientific speculation has done in the effort to explain these mysteries. Laplace's hypothesis can certainly find no standing ground either in the Orion Nebula or in those of a spiral configuration, whatever may be its situation with respect to the grand Nebula of Andromeda, or the "ring" and "planetary" nebulæ. Some other hypothesis more consonant with the appearances must be found. Among the many that have been proposed the most elaborate is the "Planetesimal Hypothesis" of Professors Chamberlin and Moulton. It is to be remarked that it applies to the spiral nebulæ distinctively, and not to an apparently chaotic mass of gas like the vast luminous cloud in Orion. The gist of the theory is that these curious objects are probably the result of close approaches to each other of two independent suns, reminding us of what was said on this subject when we were dealing with temporary stars. Of the previous history of these appulsing suns the theory gives us no account; they are simply supposed to arrive within what may be called an effective tide-producing distance, and then the drama begins. Some of the probable consequences of such an approach have been noticed in Chapter 5; let us now consider them a little more in detail.

Tides always go in couples; if there is a tide on one side of a globe there will be a corresponding tide on the other side. The cause is to be found in the law that the force of gravitation varies inversely as the square of the distance; the attraction on the nearest surface of the body exercised by another body is greater than on its center, and greater yet than on its opposite surface. If two great globes attract each other, each tends to draw the other out into an ellipsoidal figure; they must be more rigid than steel to resist this — and even then they cannot altogether resist. If they are liquid or gaseous they will yield readily to the force of distortion, the amount of which will depend upon their distance apart, for the nearer they are the greater becomes the tidal strain. If they are encrusted without and liquid or gaseous in the interior, the internal mass will strive to assume the figure demanded by the tidal force, and will, if it can, burst the restraining envelope. Now this is virtually the predicament of the body we call a sun when in the immediate presence of another body of similarly great mass. Such a body is presumably gaseous throughout, the component gases being held in a state of rigidity by the compression produced by the tremendous gravitational force of their own aggregate mass. At the surface such a body is enveloped in a shell of relatively cool matter. Now suppose a great attracting body, such as another sun, to approach near enough for the difference in its attraction on the two opposite sides of the body and on its center to become very great; the consequence will be a tidal deformation of the whole body, and it will lengthen out along the line of the gravitational pull and draw in at the sides, and if its shell offers considerable resistance, but not enough to exercise a complete restraint, it will be violently burst apart, or blown to atoms, and the internal mass will leap out on the two opposite sides in great fiery spouts. In the case of a sun further advanced in cooling than ours the interior might be composed of molten matter while the exterior crust had become rigid like the shell of an egg; then the force of the "tidal explosion" produced by the appulse of another sun would be more violent in consequence of the greater resistance overcome. Such, then, is the mechanism of the first phase in the history of a spiral nebula according to the Planetesimal Hypothesis. Two suns, perhaps extinguished ones, have drawn near together, and an explosive outburst has occurred in one or both. The second phase calls for a more agile exercise of the imagination.

To simplify the case, let us suppose that only one of the tugging suns is seriously affected by the strain. Its vast wings produced by the outburst are twisted into spirals by their rotation and the contending attractions exercised upon them, as the two suns, like battleships in desperate conflict, curve round each other, concentrating their destructive energies. Then immense quantities of débris are scattered about in which eddies are created, and finally, as the sun that caused the damage goes on its way, leaving its victim to repair its injuries as it may, the dispersed matter cools, condenses, and turns into streams of solid particles circling in elliptical paths about their parent sun. These particles, or fragments, are the "planetesimals" of the theory. In consequence of the inevitable intersection of the orbits of the planetesimals, nodes are formed where the flying particles meet, and at these nodes large masses are gradually accumulated. The larger the mass the greater its attraction, and at last the nodal points become the nuclei of great aggregations from which planets are shaped.

This, in very brief form, is the Planetesimal Hypothesis which we are asked to substitute for that based on Laplace's suggestion as an explanation of the mode of origin of the solar system; and the phenomena of the spiral nebulæ are appealed to as offering evident support to the new hypothesis. We are reminded that they are elliptical in outline, which accords with the hypothesis; that their spectra are not gaseous, which shows that they may be composed of solid particles like the planetesimals; and that their central masses present an oval form, which is what would result from the tidal effects, as just described. We also remember that some of them, like the Lord Rosse and the Andromeda nebulæ, are visually double, and in these cases we might suppose that the two masses represent the tide-burst suns that ventured into too close proximity. It may be added that the authors of the theory do not insist upon the appulse of two suns as the *only* way in which the planetesimals may have originated, but it is the only supposition that has been worked out.

But serious questions remain. It needs, for instance, but a glance at the Triangulum monster to convince the observer that it cannot be a solar system which is being evolved there, but rather a swarm of stars. Many of the detached masses are too vast to admit of the supposition that they are to be transformed into planets, in our sense of planets, and the distances of the stars which appear to have been originally ejected from the focal masses are too great to allow us to liken the assemblage that they form to a solar system. Then, too, no nodes such as the hypothesis calls for are visible. Moreover, in most of the spiral nebulæ the appearances favor the view that the supposititious encountering suns have not separated and gone each rejoicing on its way, after having inflicted the maximum possible damage on its opponent, but that, on the contrary, they remain in close association like two wrestlers who cannot escape from each other's grasp. And this is exactly what the law of gravitation demands; stars cannot approach one another with impunity, with regard either to their physical make-up or their future independence of movement. The theory undertakes to avoid this difficulty by assuming that in the case of our system the approach of the foreign body to the sun was not a close one — just close enough to produce the tidal extrusion of the relatively insignificant quantity of matter needed to form the planets. But even then the effect of the appulse would be to change the direction of flight, both of the sun and of its visitor, and there is no known star in the sky which can be selected as the sun's probable partner in their ancient *pas deux*. That there are unconquered difficulties in Laplace's hypothesis no one would deny, but in simplicity of conception it is incomparably more satisfactory, and with proper modifications could probably be made more consonant with existing facts in our solar system than that which is offered to replace it. Even as an explanation of the spiral nebulæ, not as solar systems in process of formation, but as the birthplaces of stellar clusters, the Planetesimal Hypothesis would be open to many objections. Granting its assumptions, it has undoubtedly a strong mathematical framework, but the trouble is not with the mathematics but with the assumptions. Laplace was one of the ablest mathematicians that ever lived, but he had never seen a spiral nebula; if he had, he might have invented a hypothesis to suit its phenomena. His actual hypothesis was intended only for our solar system, and he left it in the form of a "note" for the consideration of his successors, with the hope that they might be able to discover the full truth, which he confessed was hidden from him. It cannot be said that

that truth has yet been found, and when it is found the chances are that intuition and not logic will have led to it.

The spiral nebulæ, then, remain among the greatest riddles of the universe, while the gaseous nebulæ, like that of Orion, are no less mysterious, although it seems impossible to doubt that both forms give birth to stars. It is but natural to look to them for light on the question of the origin of our planetary system; but we should not forget that the scale of the phenomena in the two cases is vastly different, and the forces in operation may be equally different. A hill may have been built up by a glacier, while a mountain may be the product of volcanic forces or of the upheaval of the strata of the planet.

The Banners of the Sun

*A*s all the world knows, the sun, a blinding globe pouring forth an inconceivable quantity of light and heat, whose daily passage through the sky is caused by the earth's rotation on its axis, constitutes the most important phenomenon of terrestrial existence. Viewed with a dark glass to take off the glare, or with a telescope, its rim is seen to be a sharp and smooth circle, and nothing but dark sky is visible around it. Except for the interference of the moon, we should probably never have known that there is anymore of the sun than our eyes ordinarily see.

But when an eclipse of the sun occurs, caused by the interposition of the opaque globe of the moon, we see its immediate surroundings, which in some respects are more wonderful than the glowing central orb. These surroundings, although not in the sense in which we apply the term to the gaseous envelope of the earth, may be called the sun's atmosphere. They consist of two very different parts — first, the red "prominences," which resemble tongues of flame ascending thousands of miles above the sun's surface; and, second, the "corona," which extends to distances of millions of miles from the sun, and shines with a soft, glowing light. The two combined, when well seen, make a spectacle without parallel among the marvels of the sky. Although many attempts have been made to render the corona visible when there is no eclipse, all have failed, and it is to the moon alone that we owe its revelation. To cover the sun's disk with a circular screen will not answer the purpose because of the illumination of the air all about the observer. When the moon hides the sun, on the other hand, the sunlight is withdrawn from a great cylinder of air extending to the top of the atmosphere and spreading many miles around the observer. There is then no glare to interfere with the spectacle, and the corona appears in all its surprising beauty. The prominences, however, although they were discovered during an eclipse, can now, with the aid of the spectroscope, be seen at any time. But the prominences are rarely large enough to be noticed by the naked eye, while the streamers of the corona, stretching far away in space, like ghostly banners blown out from the black circle of the obscuring moon, attract every eye, and to this weird apparition much of the fear inspired by eclipses has been due. But if the corona has been a cause of terror in the past it has become a source of growing knowledge in our time.

The story of the first scientific observation of the corona and the prominences is thrillingly interesting, and in fact dramatic. The observation was made during the eclipse of 1842, which fortunately was visible all over Central and Southern Europe so that scores of astronomers saw it. The interest centers in what happened at Pavia in Northern Italy, where the English astronomer Francis Baily had set up his telescope. The eclipse had begun and Bailey was busy at his telescope when, to quote his own words in the account which he wrote for the *Memoirs of the Royal Astronomical Society:*

I was astounded by a tremendous burst of applause from the streets below, and at the same moment was electrified by the sight of one of the most brilliant and splendid phenomena that can well be imagined; for at that instant the dark body of the moon was suddenly surrounded with a corona, or kind of bright glory, similar in shape and magnitude to that which painters draw round the heads of saints. . .

Pavia contains many thousand inhabitants, the major part of whom were at this early hour walking about the streets and squares or looking out of windows in order to witness this long-talked-of phenomenon; and when the total obscuration took place, which was *instantaneous,* there was a universal shout from every observer which "made the welkin ring," and for the moment withdrew my attention from the object with which I was immediately occupied. I had, indeed, expected the appearance of a luminous circle round the moon during the time of total obscurity; but I did not expect, from any of the accounts of preceding eclipses that I had read, to witness so magnificent an exhibition as that which took place. . .

Splendid and astonishing, however, as this remarkable phenomenon really was, and although it could not fail to call forth the admiration and applause of every beholder, yet I must confess that there was at the same time something in its singular and wonderful appearance that was appalling. . .

But the most remarkable circumstance attending the phenomenon was the appearance of *three large protuberances* apparently emanating from the circumference of the moon, but evidently forming a portion of the corona. They had the appearance of mountains of a prodigious elevation; their color was red tinged with lilac or purple; perhaps the color of the peach-blossom would more nearly represent it. They somewhat resembled the tops of the snowy Alpine mountains when colored by the rising or the setting sun. They resembled the Alpine mountains in another respect, inasmuch as their light was perfectly steady, and had none of that flickering or sparkling motion so visible in other parts of the corona. . .

The whole of these protuberances were visible even to the last moment of total obscuration, and when the first ray of light was admitted from the sun they vanished, with the corona, altogether, and daylight was instantly restored.

I have quoted nearly all of this remarkable description not alone for its intrinsic interest, but because it is the best depiction that can be found of the general phenomena of a total solar eclipse. Still, not every such eclipse offers an equally magnificent spectacle. The eclipses of 1900 and 1905, for instance, which were seen by the writer, the first in South Carolina and the second in Spain, fell far short of that described by Bailey in splendor and impressiveness. Of course, something must be allowed for the effect of surprise; Bailey had not expected to see what was so suddenly disclosed to him. But both in 1900 and 1905 the amount of scattered light in the sky was sufficient in itself to make the corona appear faint, and there were no very conspicuous prominences visible. Yet on both occasions there was manifest among the spectators that mingling of admiration and awe of which Bailey speaks. The South Carolinians gave a cheer and the ladies waved their handkerchiefs when the corona, ineffably delicate of form and texture, *melted* into sight and then in two minutes melted away again. The Spaniards, crowded on the citadel hill of Burgos, with their king and his royal retinue in their midst, broke out with a great clapping of hands as the awaited spectacle unfolded itself in the sky; and on both occasions, before the applause began, after an awed silence a low murmur ran through the crowds. At Burgos it is said many made the sign of the cross.

It was not long before Bailey's idea that the prominences were a part of the corona was abandoned, and it was perceived that the two phenomena were to a great extent independent. At the eclipse of 1868, which the astronomers, aroused by the wonderful scene of 1842, and eager to test the powers of the newly invented spectroscope, flocked to India to witness, Janssen conceived the idea of employing the spectroscope to render the prominences visible when there was no eclipse. He succeeded the very next day, and these phenomena have been studied in that way ever since.

There are recognized two kinds of prominences — the "eruptive" and the "quiescent." The latter, which are cloud-like in form, may be seen almost anywhere along the edge of the sun; but the former, which often shoot up as if hurled from mighty volcanoes, appear to be associated with sunspots, and appear only above the zones where spots abound. Either of them, when seen in projection against the brilliant solar disk, appears white, not red, as against a background of sky. The quiescent prominences, whose elevation is often from forty thousand to sixty thousand miles, consist, as the spectroscope shows, mainly of hydrogen and helium. The latter, it will be remembered, is an element which was known to be in the sun many years before the discovery that it also exists in small quantities on the earth. A fact which may have a significance which we cannot at present see is that the emanation from radium gradually and spontaneously changes into helium, an alchemistical feat of nature that has opened many curious vistas to speculative thinkers. The eruptive prominences, which do not spread horizontally like the others, but ascend with marvelous velocity to elevations of half a million miles or more, are apparently composed largely of metallic vapors — *i.e.* metals which are usually solid on the earth, but which at solar temperatures are kept in a volatilized state. The velocity of their ascent occasionally amounts to three hundred or four hundred miles per second. It is known from mathematical considerations that the gravitation of the sun would not be able to bring back any body that started from its surface with a velocity exceeding three hundred and eighty-three miles per second; so it is evident that some of the matter hurled forth in eruptive prominences may escape from solar control and go speeding out into space, cooling and condensing into solid masses. There seems to be no reason why some of the projectiles from the sun might not reach the planets. Here, then, we have on a relatively small scale, *explosions* recalling those which it has been imagined may be the originating cause of some of the sudden phenomena of the stellar heavens.

Of the sun-spots it is not our intention here specifically to speak, but they evidently have an intimate connection with eruptive prominences, as well as some relation, not yet fully understood, with the corona. Of the real cause of sun-spots we know virtually nothing, but recent studies by Professor Hale and others have revealed a strange state of things in the clouds of metallic vapors floating above them and their surroundings. Evidences of a cyclonic tendency have been found, and Professor Hale has proved that sun-spots are strong magnetic fields, and consist of columns of ionized vapors rotating in opposite directions in the two hemispheres. A fact which may have the greatest significance is that titanium and vanadium have been found both in sun-spots and in the remarkable variable Mira Ceti, a star which every eleven months, or thereabout, flames up with great brilliancy and then sinks back to invisibility with the naked eye. It has been suggested that sun-spots are indications of the beginning of a process in the sun which will be intensified until it falls into the state of such a star as Mira. Stars very far advanced in evolution, without showing variability, also exhibit similar spectra; so that there is much reason for regarding sunspots as emblems of advancing age.

The association of the corona with sun-spots is less evident than that of the eruptive prominences; still such an association exists, for the form and extent of the corona vary with the sun-spot period of which we shall presently speak. The constitution of the corona remains to be discovered. It is evidently in part gaseous, but it also probably contains matter in the form of dust and small meteors. It includes one substance altogether mysterious — "coronium." There are reasons for thinking that this may be the lightest of all the elements, and Professor Young, its discoverer, said that it was "absolutely unique in nature; utterly distinct from any other known form of matter, terrestrial, solar, or cosmical." The enormous extent of the corona is one of its riddles. Since the development of the curious subject of the "pressure of light" it has been proposed to account for the sustentation of the corona by supposing that it is borne upon the billows of light continually poured out from the sun. Experiment has proved, what mathematical considerations had previously pointed out as probable, that the waves of light exert a pressure or driving force, which becomes evident in its effects if the body acted upon is sufficiently small. In that case the light pressure will prevail over the attraction of gravitation, and propel the attenuated matter away from the sun in the teeth of its attraction. The earth itself would be driven away if, instead of consisting of a solid globe of immense aggregate mass, it were a cloud of microscopic particles. The reason is that the pressure varies in proportion to the *surface* of the body acted upon, while the gravitational attraction is proportional to the *volume*, or the total amount of matter in the body. But the surface of any body depends upon the *square* of its diameter, while the volume depends upon the *cube* of the diameter. If, for instance, the diameter is represented by 4, the surface will be proportional to 4×4, or 16, and the volume to $4 \times 4 \times 4$, or 64; but if the diameter is taken as 2, the surface will be 2 × 2, or 4, and the volume 2 × 2 × 2, or 8. Now, the ratio of 4 to 8 is twice as great as that of 16 to 64. If the diameter is still further decreased, the ratio of the surface to the volume will proportionally grow larger; in other words, the pressure will gain upon the attraction, and whatever their original ratio may have been, a time will come, if the diminution of size continues, when the pressure will become more effective than the attraction, and the body will be driven away. Supposing the particles of the corona to be below the critical size for the attraction of a mass like that of the sun to control them, they would be driven off into the surrounding space and appear around the sun like the clouds of dust around a mill. We shall return to this subject in connection with the Zodiacal Light, the Aurora, and Comets.

On the other hand, there are parts of the corona which suggest by their forms the play of electric or magnetic forces. This is beautifully shown in some of the photographs that have been made of the corona during recent eclipses. Take, for instance, that of the eclipse of 1900. The sheaves of light emanating from the poles look precisely like the "lines of force" surrounding the poles of a magnet. It will be noticed in this photograph that the corona appears to consist of two portions: one comprising the polar rays just spoken of, and the other consisting of the broader, longer, and less-defined masses of light extending out from the equatorial and middle-latitude zones. Yet even in this more diffuse part of the phenomenon one can detect the presence of submerged curves bearing more or less resemblance to those about the poles. Just what part electricity or electro-magnetism plays in the mechanism of the solar radiation it is impossible to say, but on the assumption that it is a very important part is based the hypothesis that there exists a direct solar influence not only upon the magnetism, but upon the weather of the earth. This hypothesis has been under discussion for half a century, and still we do not know just how much truth it represents. It is certain that the outbreak of great disturbances on the sun, accompanied by the formation of sun-spots and the upshooting of eruptive prominences (phenomena which we should naturally expect to be attended by action), have been instantly followed by corresponding "magnetic storms" on the earth and brilliant displays of the auroral lights. There have been occasions when the influence has manifested itself in the most startling ways, a great solar outburst being followed by a mysterious gripping of the cable and telegraph systems of the world, as if an invisible and irresistible hand had seized them. Messages are abruptly cut off, sparks leap from the telegraph instruments, and the entire earth seems to have been thrown into a magnetic flurry. These occurrences affect the mind with a deep impression of the dependence of our planet on the sun, such as we do not derive from the more familiar action of the sunlight on the growth of plants and other phenomena of life depending on solar influences.

Perhaps the theory of solar magnetic influence upon the weather is best known in connection with the "sun-spot cycle." This, at any rate, is, as already remarked, closely associated with the corona. Its existence was discovered in 1843 by the German astronomer Schwabe. It is a period of variable length, averaging about eleven years, during which the number of spots visible on the sun first increases to a maximum, then diminishes to a minimum, and finally increases again to a maximum. For unknown reasons the period is sometimes two or three years longer than the average and sometimes as much shorter. Nevertheless, the phenomena always recur in the same order. Starting, for instance, with a time when the observer can find few or no spots, they gradually increase in number and size until a maximum, in both senses, is reached, during which the spots are often of enormous size and exceedingly active. After two or three years they begin to diminish in number, magnitude, and activity until they almost or quite disappear. A strange fact is that when a new period opens, the spots appear first in high northern and southern latitudes, far from the solar equator, and as the period advances they not only increase in number and size, but break out nearer and nearer to the equator, the last spots of a vanishing period sometimes lingering in the equatorial region after the advance-guard of its successor has made its appearance in the high latitudes. Spots are never seen on the equator nor near the poles. It was not very long after the discovery of the sun-spot cycle that the curious observation was made that a striking coincidence existed between the period of the sun-spots and another period affecting the general magnetic condition of the earth. When a curved line representing the varying number of sun-spots was compared with another curve showing the variations in the magnetic state of the earth the two were seen to be in almost exact accord, a rise in one curve corresponding to a rise in the other, and a fall to a fall. Continued observation has proved that this is a real coincidence and not an accidental one, so that the connection, although as yet unexplained, is accepted as established. But does the influence extend further, and directly affect the weather and the seasons as well as the magnetic elements of the earth? A final answer to this question cannot yet be given, for the evidence is contradictory, and the interpretations put upon it depend largely on the predilections of the judges.

But, in a broad sense, the sun-spots and the phenomena connected with them *must* have a relation to terrestrial meteorology, for they prove the sun to be a variable star. Reference was made, a few lines above, to the resemblance of the spectra of sun-spots to those of certain stars which seem to be failing through age. This in itself is extremely suggestive; but if this resemblance had never been discovered, we should have been justified in regarding the sun as variable in its output of energy; and not only variable, but probably increasingly so. The very inequalities in the sun-spot cycle are suspicious. When the sun is most spotted its total light may be reduced by one-thousandth part, although it is by no means certain that its outgiving of thermal radiations is then reduced. A loss of one-thousandth of its luminosity would correspond to a decrease of .0025 of a stellar magnitude, considering the sun as a star viewed from distant space. So slight a change would not be perceptible; but it is not alone sun-spots which obscure the solar surface, its entire globe is enveloped with an obscuring veil. When studied with a powerful telescope the sun's surface is seen to be thickly mottled with relatively obscure specks, so numerous that it has been estimated that they cut off from one-tenth to one-twentieth of the light that we should receive from it if the whole surface were as brilliant as its brightest parts. The condition of other stars warrants the conclusion that this obscuring envelope is the product of a process of refrigeration which will gradually make the sun more and more variable until its history ends in extinction. Looking backward, we see a time when the sun must have been more brilliant than it is now. At that time it probably shone with the blinding white splendor of such stars as Sirius, Spica, and Vega; now it resembles the relatively dull Procyon; in time it will turn ruddy and fall into the closing cycle represented by Antares. Considering that once it must have been more radiantly powerful than at present, one is tempted to wonder if that could have been the time when tropical life flourished within the earth's polar circles, sustained by a vivific energy in the sun which it has now lost.

The corona, as we have said, varies with the sun-spot cycle. When the spots are abundant and active the corona rises strong above the spotted zones, forming immense beams or streamers, which on one occasion, at least, had an observed length of *ten million miles*. At the time of a spot minimum the corona is less brilliant and has a different outline. It is then that the curved polar rays are most conspicuous. Thus the vast banners of the sun, shaken out in the eclipse, are signals to tell of its varying state, but it will probably be long before we can read correctly their messages.

The Zodiacal Light Mystery

*T*here is a singular phenomenon in the sky — one of the most puzzling of all — which has long arrested the attention of astronomers, defying their efforts at explanation, but which probably not one in a hundred, and possibly not one in a thousand, of the readers of this book has ever seen. Yet its name is often spoken, and it is a conspicuous object if one knows when and where to look for it, and when well seen it exhibits a mystical beauty which at the same time charms and awes the beholder. It is called "The Zodiacal Light," because it lies within the broad circle of the Zodiac, marking the sun's apparent annual path through the stars. What it is nobody has yet been able to find out with certainty, and books on astronomy usually speak of it with singular reserve. But it has given rise to many remarkable theories, and a true explanation of it would probably throw light on a great many other celestial mysteries. The Milky Way is a more wonderful object to look upon, but its nature can be comprehended, while there is a sort of uncanniness about the Zodiacal Light which immediately impresses one upon seeing it, for its part in the great scheme of extra-terrestrial affairs is not evident.

If you are out-of-doors soon after sunset — say, on an evening late in the month of February — you may perceive, just after the angry flush of the dying winter's day has faded from the sky, a pale ghostly presence rising above the place where the sun went down. The writer remembers from boyhood the first time it was pointed out to him and the unearthly impression that it made, so that he afterward avoided being out alone at night, fearful of seeing the spectral thing again. The phenomenon brightens slowly with the fading of the twilight, and soon distinctly assumes the shape of an elongated pyramid of pearly light, leaning toward the south if the place of observation is in the northern hemisphere. It does not impress the observer at all in the same manner as the Milky Way; that looks far off and is clearly among the stars, but the Zodiacal Light seems closer at hand, as if it were something more intimately concerning the earth. To all it immediately suggests a connection, also, with the sunken sun. If the night is clear and the moon absent (and if you are in the country, for city lights ruin the spectacles of the sky), you will be able to watch the apparition for a long time. You will observe that the light is brightest near the horizon, gradually fading as the pyramidal beam mounts higher, but in favorable circumstances it may be traced nearly to the meridian south of the zenith, where its apex at last vanishes in the starlight. It continues visible during the evenings of March and part of April, after which, ordinarily, it is seen no more, or if seen is relatively faint and unimpressive. But when autumn comes it appears again, this time not like a wraith hovering above the westward tomb of the daygod, but rather like a spirit of the morning announcing his reincarnation in the east.

The reason why the Zodiacal Light is best seen in our latitudes at the periods just mentioned is because at those times the Zodiac is more nearly perpendicular to the horizon, first in the west and then in the east; and, since the phenomenon is confined within the borders of the Zodiac, it cannot be favorably placed for observation when the zodiacal plane is but slightly inclined to the horizon. Its faint light requires the contrast of a background of dark sky in order to be readily perceptible. But within the tropics, where the Zodiac is always at a favorable angle, the mysterious light is more constantly visible. Nearly all observant travelers in the equatorial regions have taken particular note of this phenomenon, for being so much more conspicuous there than in the temperate zones it at once catches the eye and holds the attention as a novelty. Humboldt mentions it many times in his works, for his genius was always attracted by things out of the ordinary and difficult of explanation, and he made many careful observations on its shape, its brilliancy, and its variations; for there can be no doubt that it does vary, and sometimes to an astonishing degree. It is said that it once remained practically invisible in Europe for several years in succession. During a trip to South Africa in 1909 an English astronomer, Mr. E.W. Maunder, found a remarkable difference between the appearance of the Zodiacal Light on his going and coming voyages. In fact, when crossing the equator going south he did not see it at all; but on returning he had, on March 6th, when one degree south of the equator, a memorable view of it.

> It was a bright, clear night, and the Zodiacal Light was extraordinarily brilliant — brighter than he had ever seen it before. The Milky Way was not to be compared with it. The brightest part extended 75° from the sun. There was a faint and much narrower extension which they could just make out beyond the Pleiades along the ecliptic, but the greater part of the Zodiacal Light showed as a broad truncated column, and it did not appear nearly as conical as he had before seen it.

When out of the brief twilight of intertropical lands, where the sun drops vertically to the horizon and night rushes on like a wave of darkness, the Zodiacal Light shoots to the very zenith, its color is described as a golden tint, entirely different from the silvery sheen of the Milky Way. If I may venture again to refer to personal experiences and impressions, I will recall a view of the Zodiacal Light from the summit of the cone of Mt Etna in the autumn of the year 1896 (more briefly described in *Astronomy with the Naked Eye*). There are few lofty mountains so favorably placed as Etna for observations of this kind. It was once resorted to by Prof. George E. Hale, in an attempt to see the solar corona without an eclipse. Rising directly from sea level to an elevation of nearly eleven thousand feet, the observer on its summit at night finds himself, as it were, lost in the midst of the sky. But for the black flanks of the great cone on which he stands he might fancy himself to be in a balloon. On the occasion to which I refer the world beneath was virtually invisible in the moonless night. The blaze of the constellations overhead was astonishingly brilliant, yet amid all their magnificence my attention was immediately drawn to a great tapering light that sprang from the place on the horizon where the sun would rise later, and that seemed to be blown out over the stars like a long, luminous veil. It was the finest view of the Zodiacal light that I had ever enjoyed — thrilling in its strangeness — but I was almost disheartened by the indifference of my guide, to whom it was only a light and nothing more. If he had no science, he had less poetry — rather a remarkable thing, I thought, for a child of his clime. The Light appeared to me to be distinctly brighter than the visible part of the Milky Way which included the brilliant stretches in Auriga and Perseus, and its color, if one may speak of color in connection with such an object, seemed richer than that of the galactic band; but I did not think of it as yellow, although Humboldt has described it as resembling a golden curtain drawn over the stars, and Du Chaillu in Equatorial Africa found it of a bright yellow color. It may vary in color as in conspicuousness. The fascination of that extraordinary sight has never faded from my memory. I turned to regard it again and again, although I had never seen the stellar heavens so brilliant, and it was one of the last things I looked for when the morning glow began softly to mount in the east, and Sicily and the Mediterranean slowly emerged from the profound shadow beneath us.

The Zodiacal Light seems never to have attracted from astronomers in general the amount of careful attention that it deserves; perhaps because so little can really be made of it as far as explanation is concerned. I have referred to the restraint that scientific writers apparently feel in speaking of it. The grounds for speculation that it affords may be too scanty to lead to long discussions, yet it piques curiosity, and as we shall see in a moment has finally led to a most interesting theory. Once it was the subject of an elaborate series of studies which carried the observer all round the world. That was in 1845 — 46, during the United States Exploring Expedition that visited the then little known Japan. The chaplain of the fleet, the Rev. Mr. Jones, went out prepared to study the mysterious light in all its phases. He saw it from many latitudes on both sides of the equator, and the imagination cannot but follow him with keen interest in his world-circling tour, keeping his eyes every night fixed upon the phantasm overhead, whose position shifted with that of the hidden sun. He demonstrated that the flow extends at times completely across the celestial dome, although it is relatively faint directly behind the earth. On his return the government published a large volume of his observations, in which he undertook to show that the phenomenon was due to the reflection of sunlight from a ring of meteoric bodies encircling the earth. But, after all, this elaborate investigation settled nothing.

Prof. E.E. Barnard has more recently devoted much attention to the Zodiacal Light, as well as to a strange attendant phenomenon called the "Gegenschein," or Counterglow, because it always appears at that point in the sky which is exactly opposite the sun. The Gegenschein is an extremely elusive phenomenon, suitable only for eyes that have been specially trained to see it. Professor Newcomb has cautiously remarked that

> it is said that in that point of the heavens directly opposite the sun there is an elliptical patch of light. . . This phenomenon is so difficult to account for that its existence is sometimes doubted; yet the testimony in its favor is difficult to set aside.

It certainly cannot be set aside at all since the observations of Barnard. I recall an attempt to see it under his guidance during a visit to Mount Hamilton, when he was occupied there with the Lick telescope. Of course, both the Gegenschein and the Zodiacal Light are too diffuse to be studied with telescopes, which, so to speak, magnify them out of existence. They can only be successfully studied with the naked eye, since every faintest glimmer that they afford must be utilized. This is especially true of the Gegenschein. At Mount Hamilton, Mr. Barnard pointed out to me its location with reference to certain stars, but with all my gazing I could not be sure that I saw it. To him, on the contrary, it was obvious; he had studied it for months, and was able to indicate its shape, its boundaries, its diameter, and the declination of its center with regard to the ecliptic. There is not, of course, the shadow of a doubt of the existence of the Gegenschein, and yet I question if one person in a million has ever seen or ever will see it. The Zodiacal Light, on the other hand, is plain enough, provided that the time and the circumstances of the observation are properly chosen.

In the attempts to explain the Zodiacal Light, the favorite hypothesis has been that it is an appendage of the sun — perhaps simply an extension of the corona in the plane of the ecliptic, which is not very far from coinciding with that of the sun's equator. This idea is quite a natural one, because of the evident relation of the light to the position of the sun. The vast extension of the equatorial wings of the corona in 1878 gave apparent support to this hypothesis; if the substance of the corona could extend ten million miles from the sun, why might it not extend even one hundred million, gradually fading out beyond the orbit of the earth? A variation of this hypothesis assumes that the reflection is due to swarms of meteors circling about the sun, in the plane of its equator, all the way from its immediate neighborhood to a distance exceeding that of the earth. But in neither form is the hypothesis satisfactory; there is nothing in the appearance of the corona to indicate that it extends even as far as the planet Mercury, while as to meteors, the orbits of the known swarms do not accord with the hypothesis, and we have no reason to believe that clouds of others exist traveling in the part of space where they would have to be in order to answer the requirements of the theory. The extension of the corona in 1878 did not resemble in its texture the Zodiacal Light.

Now, it has so often happened in the history of science that an important discovery in one branch has thrown unexpected but most welcome light upon some pending problem in some other branch, that a strong argument might be based upon that fact alone against the too exclusive devotion of many investigators to the narrow lines of their own particular specialty; and the Zodiacal Light affords a case in point, when it is considered in connection with recent discoveries in chemistry and physics. From the fact that atoms are compound bodies made up of corpuscles at least a thousand times smaller than the smallest known atom — a fact which astounded most men of science when it was announced a few years ago — a new hypothesis has been developed concerning the nature of the Zodiacal Light (as well as other astronomical riddles), and this hypothesis comes not from an astronomer, but from a chemist and physicist, the Swede, Svante Arrhenius. In considering an outline of this new hypothesis we need neither accept nor reject it; it is a case rather for suspension of judgment.

To begin with, it carries us back to the "pressure of light" mentioned in the preceding chapter. The manner in which this pressure is believed generally to act was there sufficiently explained, and it only remains to see how it is theoretically extended to the particles of matter supposed to constitute the Zodiacal Light. We know that corpuscles, or "fragments of atoms" negatively electrified, are discharged from hot bodies. Streams of these "ions" pour from many flames and from molten metals; and the impact of the cathode and ultra-violet rays causes them to gush even from cold bodies. In the vast laboratory of the sun it is but reasonable to suppose that similar processes are taking place. "As a very hot metal emits these corpuscles," says Prof. J.J. Thomson, "it does not seem an improbable hypothesis that they are emitted by that very hot body, the sun." Let it be assumed, then, that the sun does emit them; what happens next? Negatively charged corpuscles, it is known, serve as nuclei to which particles of matter in the ordinary state are attracted, and it is probable that those emitted from the sun immediately pick up loads in this manner and so grow in bulk. If they grow large enough the gravitation of the sun draws them back, and they produce a negative charge in the solar atmosphere. But it is probable that many of the particles do not attain the critical size which, according to the principles before explained, would enable the gravitation of the sun to retain them in opposition to the pressure of the waves of light, and with these particles the light pressure is dominant. Clouds of them may be supposed to be continually swept away from the sun into surrounding space, moving mostly in or near the plane of the solar equator, where the greatest activity, as indicated by sunspots and related phenomena, is taking place. As they pass outward into space many of them encounter the earth. If the earth, like the moon, had no atmosphere the particles would impinge directly on its surface, giving it a negative electric charge. But the presence of the atmosphere changes all that, for the first of the flying particles that encounter it impart to it their negative electricity, and then, since like electric charges repel like, the storm of particles following will be sheered off from the earth, and will stream around it in a maze of hyperbolic paths. Those that continue on into space beyond the earth may be expected to continue picking up wandering particles of matter until their bulk has become so great that the solar attraction prevails again over the light pressure acting upon them, and they turn again sunward. Passing the earth on their return they will increase the amount of dust-clouds careering round it; and these will be further increased by the action of the ultra-violet rays of the sunlight causing particles to shoot radially away from the earth when the negative charge of the upper atmosphere has reached a certain amount, which particles, although starting sunward, will be swept back to the earth with the oncoming streams. As the final result of all this accumulation of flying and gyrating particles in the earth's neighborhood, we are told that the latter must be transformed into the semblance of a gigantic solid-headed comet provided with streaming tails, the longest of them stretching away from the direction of the sun, while another shorter one extends toward the sun. This shorter tail is due to the particles that we have just spoken of as being driven sunward from the earth by the action of ultra-violet light. No doubt this whole subject is too technical for popular statement; but at any rate the general reader can understand the picturesque side of the theory, for its advocates assure us that if we were on the moon we would doubtless be able to see the cometlike tails of the earth, and then we could appreciate the part that they play in producing the phenomenon of the Zodiacal Light.

That the Light as we see it could be produced by the reflection of sunlight from swarms of particles careering round the earth in the manner supposed by Arrhenius' hypothesis is evident enough; and it will be observed that the new theory, after all, is only another variant of the older one which attributes the Zodiacal Light to an extension of the solar corona. But it differs from the older theory in offering an explanation of the manner in which the extension is effected, and it differentiates between the corona proper and the streams of negative particles shot away from the sun. In its details the hypothesis of Arrhenius also affords an explanation of many peculiarities of the Zodiacal Light, such as that it is confined to the neighborhood of the ecliptic, and that it is stronger on the side of the earth which is just turning away from a position under the sun than on the other side; but it would carry us beyond our limits to go into these particulars. The Gegenschein, according to this theory, is a part of the same phenomenon as the Zodiacal Light, for by the laws of perspective it is evident that the reflection from the streams of particles situated at a point directly opposite to the sun would be at a maximum, and this is the place which the Gegenschein occupies. Apart from its geometrical relations to the position of the sun, the variability of the Zodiacal Light appears to affirm its solar dependence, and this too would be accounted for by Arrhenius' hypothesis better than by the old theory of coronal extension. The amount of corpuscular discharge from the sun must naturally be governed by the state of relative activity or inactivity of the latter, and this could not but be reflected in the varying splendor of the Zodiacal Light. But much more extended study than has yet been given to the subject will be required before we can feel that we know with reasonable certainty what this mysterious phenomenon really is. By the hypothesis of Arrhenius every planet that has an atmosphere must have a Zodiacal Light attending it, but the phenomenon is too faint for us to be able to see it in the case, for instance, of Venus, whose atmosphere is very abundant. The moon has no corresponding "comet's tail" because, as already explained, of the lack of a lunar atmosphere to repel the streams by becoming itself electrified; but if there were a lunar Zodiacal Light, no doubt we could see it because of the relative nearness of our satellite.

Marvels of the Aurora

One of the most vivid recollections of my early boyhood is that of seeing my father return hastily into the house one evening and call out to the family: "Come outside and look at the sky!" Ours was a country house situated on a commanding site, and as we all emerged from the doorway we were dumbfounded to see the heavens filled with pale flames which ran licking and quivering over the stars. Instantly there sprang into my terrified mind the recollection of an awful description of "the Day of Judgment" (the *Dies Iræ),* which I had heard with much perturbation of spirit in the Dutch Reformed church from the lips of a tall, dark-browed, dreadfully-in-earnest preacher of the old-fashioned type. My heart literally sank at sight of the spectacle, for it recalled the preacher's very words; it was just as he had said it would be, and it needed the assured bearing of my elders finally to convince me that

> That Day of Wrath, O dreadful day,
> When Heaven and Earth shall pass away,
> As David and the Sibyl say

had not actually come upon us. And even the older members of the household were not untouched with misgivings when menacing spots of crimson appeared, breaking out now here, now there, in the shuddering sky. Toward the north the spectacle was appalling. A huge arch spanned an unnaturally dark segment resting on the horizon, and above this arch sprang up beams and streamers in a state of incessant agitation, sometimes shooting up to the zenith with a velocity that took one's breath, and sometimes suddenly falling into long ranks, and *marching, marching, marching,* like an endless phalanx of fiery specters, and moving, as I remember, always from east to west. The absolute silence with which these mysterious evolutions were performed and the quavering reflections which were thrown upon the ground increased the awfulness of the exhibition. Occasionally enormous curtains of lambent flame rolled and unrolled with a majestic motion, or were shaken to and fro as if by a mighty, noiseless wind. At times, too, a sudden billowing rush would be made toward the zenith, and for a minute the sky overhead would glow so brightly that the stars seemed to have been consumed. The spectacle continued with varying intensity for hours.

This exhibition occurred in Central New York, a latitude in which the Aurora Borealis is seldom seen with so much splendor. I remember another similar one seen from the city of New York in November, 1882. On this last occasion some observers saw a great upright beam of light which majestically moved across the heavens, stalking like an apparition in the midst of the auroral pageant, of whose general movements it seemed to be independent, maintaining always its upright posture, and following a magnetic parallel from east to west. This mysterious beam was seen by no less than twenty-six observers in different parts of the country, and a comparison of their observations led to a curious calculation indicating that the apparition was about *one hundred and thirty-three miles tall* and moved at the speed of ten miles per second!

But, as everybody knows, it is in the Arctic regions that the Aurora, or the "Northern Lights," can best be seen. There, in the long polar night, when for months together the sun does not rise, the strange coruscations in the sky often afford a kind of spectral daylight in unison with the weird scenery of the world of ice. The pages in the narratives of Arctic exploration that are devoted to descriptions of the wonderful effects of the Northern Lights are second to none that man has ever penned in their fascination. The lights, as I have already intimated, display astonishing colors, particularly shades of red and green, as they flit from place to place in the sky. The discovery that the magnetic needle is affected by the

Aurora, quivering and darting about in a state of extraordinary excitement when the lights are playing in the sky, only added to the mystery of the phenomenon until its electromagnetic nature had been established. This became evident as soon as it was known that the focus of the displays was the magnetic pole; and when the far South was visited the Aurora Australis was found, having its center at the South Magnetic Pole. Then, if not before, it was clear that the earth was a great globular magnet, having its poles of opposite magnetism, and that the auroral lights, whatever their precise cause might be, were manifestations of the magnetic activity of our planet. After the invention of magnetic telegraphy it was found that whenever a great Aurora occurred the telegraph lines were interrupted in their operation, and the ocean cables ceased to work. Such a phenomenon is called a "magnetic storm."

The interest excited by the Aurora in scientific circles was greatly stimulated when, in the last half of the nineteenth century, it was discovered that it is a phenomenon intimately associated with disturbances on the sun. The ancient "Zurich Chronicles," extending from the year 1000 to the year 1800, in which both sun-spots visible to the naked eye and great displays of the auroral lights were recorded, first set Rudolf Wolf on the track of this discovery. The first notable proof of the suspected connection was furnished with dramatic emphasis by an occurrence which happened on September 1, 1859. Near noon on that day two intensely brilliant points suddenly broke out in a group of sun-spots which were under observation by Mr. R.C. Carrington at his observatory at Redhill, England. The points remained visible for not more than five minutes, during which interval they moved *thirty-five thousand miles* across the solar disk. Mr. R. Hodgson happened to see the same phenomenon at his observatory at Highgate, and thus all possibility of deception was removed. But neither of the startled observers could have anticipated what was to follow, and, indeed, it was an occurrence which has never been precisely duplicated. I quote the eloquent account given by Miss Clerke in her *History of Astronomy During the Nineteenth Century.*

This unique phenomenon seemed as if specially designed to accentuate the inference of a sympathetic relation between the earth and the sun. From August 28 to September 4, 1859, a magnetic storm of unparalleled intensity, extent, and duration was in progress over the entire globe. Telegraphic communication was everywhere interrupted — except, indeed, that it was in some cases found practicable to work the lines *without batteries* by the agency of the earth-currents alone; sparks issued from the wires; gorgeous auroras draped the skies in solemn crimson over both hemispheres, and even in the tropics; the magnetic needle lost all trace of continuity in its movements and darted to and fro as if stricken with inexplicable panic. The coincidence was even closer. *At the very instant* of the solar outburst witnessed by Carrington and Hodgson the photographic apparatus at Kew registered a marked disturbance of all the three magnetic elements; while shortly after the ensuing midnight the electric agitation culminated, thrilling the whole earth with subtle vibrations, and lighting up the atmosphere from pole to pole with coruscating splendors which perhaps dimly recall the times when our ancient planet itself shone as a star.

If this amazing occurrence stood alone, and as I have already said it has never been exactly duplicated, doubt might be felt concerning some of the inferences drawn from it; but in varying forms it has been repeated many times, so that now hardly anyone questions the reality of the assumed connection between solar outbursts and magnetic storms accompanied by auroral displays on the earth. It is true that the late Lord Kelvin raised difficulties in the way of the hypothesis of a direct magnetic action of the sun upon the earth, because it seemed to him that an inadmissible quantity of energy was demanded to account for such action. But no calculation like that which he made is final, since all calculations depend upon the validity of the data; and no authority is unshakable in science, because no man can possess omniscience. It was Lord Kelvin who, but a few years before the thing was actually accomplished, declared that aërial navigation was an impracticable dream, and demonstrated its impracticability by calculation. However the connection may be brought about, it is as certain as evidence can make it that solar outbursts are coincident with terrestrial magnetic disturbances, and coincident in such a way as to make the inference of a causal connection irresistible. The sun is only a little more than a hundred times its own diameter away from the earth. Why, then, with the subtle connection between them afforded by the ether which conveys to us the blinding solar light and the life-sustaining solar heat, should it be so difficult to believe that the sun's enormous electric energies find a way to us also? No doubt the impulse coming from the sun acts upon the earth after the manner of a touch upon a trigger, releasing energies which are already stored up in our planet.

But besides the evidence afforded by such occurrences as have been related of an intimate connection between solar outbreaks and terrestrial magnetic flurries, attended by magnificent auroral displays, there is another line of proof pointing in the same direction. Thus, it is known that the sun-spot period, as remarked in a preceding chapter, coincides in a most remarkable manner with the periodic fluctuations in the magnetic state of the earth. This coincidence runs into the most astonishing details. For instance, when the sun-spot period shortens, the auroral period shortens to precisely the same extent; as the short sun-spot periods usually bring the most intense outbreaks of solar activity, so the corresponding short auroral periods are attended by the most violent magnetic storms; a secular period of about two hundred and twenty-two years affecting sun-spots is said to have its auroral duplicate; a shorter period of fifty-five and a half years, which some observers believe that they have discovered appears also to be common to the two phenomena; and yet another "superposed" period of about thirty-five years, which some investigators aver exists, affects sun-spots and aurora alike. In short, the coincidences are so numerous and significant that one would have to throw the doctrine of probability to the winds in order to be able to reject the conclusion to which they so plainly lead.

But still the question recurs: How is the influence transmitted? Here Arrhenius comes once more with his hypothesis of negative corpuscles, or ions, driven away from the sun by light-pressure — a hypothesis which seems to explain so many things — and offers it also as an explanation of the way in which the sun creates the Aurora. He would give the Aurora the same lineage with the Zodiacal Light. To understand the application of this theory we must first recall the fact that the earth is a great magnet having its two opposite poles of magnetism, one near the Arctic and the other near the Antarctic Circle. Like all magnets, the earth is surrounded with "lines of force," which, after the manner of the curved rays we saw in the photograph of a solar eclipse, start from a pole, rising at first nearly vertically, then bend gradually over, passing high above the equator, and finally descending in converging sheaves to the opposite pole. Now the axis of the earth is so placed in space that it lies at nearly a right angle to the direction of the sun, and as the streams of negatively charged particles come pouring on from the sun (see the last preceding chapter), they arrive in the greatest numbers over the earth's equatorial regions. There they encounter the lines of magnetic force at the place where the latter have their greatest elevation above the earth, and where their direction is horizontal to the earth's surface. Obeying a law which has been demonstrated in the laboratory, the particles then follow the lines of force toward the poles. While they are above the equatorial regions they do not become luminescent, because at the great elevation that they there occupy there is virtually no atmosphere; but as they pass on toward the north and the south they begin to descend with the lines of force, curving down to meet at the poles; and, encountering a part of the atmosphere comparable in density with what remains in an exhausted Crookes tube, they produce a glow of cathode rays. This glow is conceived to represent the Aurora, which may consequently be likened to a gigantic exhibition of vacuum-tube lights. Anybody who recalls his student days in the college laboratory and who has witnessed a display of Northern Lights will at once recognize the resemblance between them in colors, forms, and behavior. This resemblance had often been noted before Arrhenius elaborated his hypothesis.

Without intending to treat his interesting theory as more than a possibly correct explanation of the phenomena of the Aurora, we may call attention to some apparently confirmatory facts. One of the most striking of these relates to a seasonal variation in the average number of auroræ. It has been observed that there are more in March and September than at any other time of the year, and fewer in June and December; moreover (and this is a delicate test as applied to the theory), they are slightly rarer in June than in December. Now all these facts seem to find a ready explanation in the hypothesis of Arrhenius, thus: (1) The particles issuing from the sun are supposed to come principally from the regions whose excitement is indicated by the presence of sun-spots (which accords with Hale's observation that sun-spots are columns of ionized vapors), and these regions have a definite location on either side of the solar equator, seldom approaching it nearer than within 5° or 10° north or south, and never extending much beyond 35° toward either pole; (2) The equator of the sun is inclined about 7° to the plane of the earth's orbit, from which it results that twice in a year — *viz.,* in June and December — the earth is directly over the solar equator, and twice a year — *viz.,* in March and September — when it is farthest north or south of the solar equator, it is over the inner edge of the sun-spot belts. Since the corpuscles must be supposed to be propelled radially from the sun, few will reach the earth when the latter is over the solar equator in June and December, but when it is over, or nearly over, the spot belts, in March and September, it will be in the line of fire of the more active parts of the solar surface, and relatively rich streams of particles will reach it. This, as will be seen from what has been said above, is in strict accord with the observed variations in the frequency of auroræ. Even the fact that somewhat fewer auroræ are seen in June than in December also finds its explanation in the known fact that the earth is about three million miles nearer the sun in the winter than in the summer, and the number of particles reaching it will vary, like the intensity of light, inversely as the square of the distance. These coincidences are certainly very striking, and they have a cumulative force. If we accept the theory, it would appear that we ought to congratulate ourselves that the inclination of the sun's equator is so slight, for as things stand the earth is never directly over the most active regions of the sun-spots, and consequently never suffers from the maximum bombardment of charged particles of which the sun is capable. Incessant auroral displays, with their undulating draperies, flitting colors, and marching columns might not be objectionable from the point of view of picturesqueness, but one magnetic storm of extreme intensity following closely upon the heels of another, for months on end, crazing the magnetic needle and continually putting the telegraph and cable lines out of commission, to say nothing of their effect upon "wireless telegraphy," would hardly add to the charms of terrestrial existence.

One or two other curious points in connection with Arrhenius' hypothesis may be mentioned. First, the number of auroræ, according to his explanation, ought to be greatest in the daytime, when the face of the earth on the sunward side is directly exposed to the atomic bombardment. Of course visual observation can give us no information about this, since the light of the Aurora is never sufficiently intense to be visible in the presence of daylight, but the records of the magnetic observatories can be, and have been, appealed to for information, and they indicate that the facts actually accord with the theory. Behind the veil of sunlight in the middle of the afternoon, there is good reason to believe, auroral exhibitions often take place which would eclipse in magnificence those seen at night if we could behold them. Observation shows, too, that auroræ are more frequent before than after midnight, which is just what we should expect if they originate in the way that Arrhenius supposes. Second, the theory offers an explanation of the alleged fact that the formation of clouds in the upper air is more frequent in years when auroræ are most abundant, because clouds are the result of the condensation of moisture upon floating particles in the atmosphere (in an absolutely dustless atmosphere there would be no clouds), and it has been proved that negative ions like those supposed to come from the sun play a master part in the phenomena of cloud formation.

Yet another singular fact, almost mystical in its suggestions, may be mentioned. It seems that the dance of the auroral lights occurs most frequently during the absence of the moon from the hemisphere in which they appear, and that they flee, in greater part, to the opposite hemisphere when the moon's revolution in an orbit considerably inclined to the earth's equator brings her into that where they have been performing.

Arrhenius himself discovered this curious relation of auroral frequency to the position of the moon north or south of the equator, and he explains it in this way. The moon, like the earth, is exposed to the influx of the ions from the sun; but having no atmosphere, or almost none, to interfere with them, they descend directly upon her surface and charge her with an electric negative potential to a very high degree. In consequence of this she affects the electric state of the upper parts of the earth's atmosphere where they lie most directly beneath her, and thus prevents, to a large extent, the negative discharges to which the appearance of the Aurora is due. And so "the extravagant and erring spirit" of the Aurora avoids the moon as Hamlet's ghost fled at the voice of the cock announcing the awakening of the god of day.

There are even other apparent confirmations of the hypothesis, but we need not go into them. We shall, however, find one more application of it in the next chapter, for it appears to be a kind of cure-all for astronomical troubles; at any rate it offers a conceivable solution of the question, How does the sun manage to transmit its electric influence to the earth? And this solution is so grandiose in conception, and so novel in the mental pictures that it offers, that its acceptance would not in the least detract from the impression that the Aurora makes upon the imagination.

Strange Adventures of Comets

The fears and legends of ancient times before Science was born, and the superstitions of the Dark Ages, sedulously cultivated for theological purposes by monks and priests, have so colored our ideas of the influence that comets have had upon the human mind that many readers may be surprised to learn that it was the apparition of a wonderful comet, that of 1843, which led to the foundation of our greatest astronomical institution, the Harvard College Observatory. No doubt the comet superstition existed half a century ago, as, indeed, it exists yet today, but in this case the marvelous spectacle in the sky proved less effective in inspiring terror than in awakening a desire for knowledge. Even in the sixteenth century the views that enlightened minds took of comets tended powerfully to inspire popular confidence in science, and Halley's prediction, after seeing and studying the motion of the comet which appeared in 1682, that it would prove to be a regular member of the sun's family and would be seen returning after a period of about seventy-six years, together with the fulfillment of that prediction, produced a revulsion from the superstitious notions which had so long prevailed.

Then the facts were made plain that comets are subject to the law of gravitation equally with the planets; that there are many which regularly return to the neighborhood of the sun (perihelion); and that these travel in orbits differing from those of the planets only in their greater eccentricity, although they have the peculiarity that they do not, like the planets, all go round the sun in the same direction, and do not keep within the general plane of the planetary system, but traverse it sometimes from above and sometimes from below. Other comets, including most of the "great" ones, appear to travel in parabolic or, in a few cases, hyperbolic orbits, which, not being closed curves, never bring them back again. But it is not certain that these orbits may not be extremely eccentric ellipses, and that after the lapse of hundreds, or thousands, of years the comets that follow them may not reappear. The question is an interesting one, because if all orbits are really ellipses, then all comets must be permanent members of the solar system, while in the contrary case many of them are simply visitors, seen once and never to be seen again. The hypothesis that comets are originally interlopers might seem to derive some support from the fact that the certainly periodic ones are associated, in groups, with the great outer planets, whose attraction appears to have served as a trap for them by turning them into elliptical orbits and thus making them prisoners in the solar system. Jupiter, owing to his great mass and his commanding situation in the system, is the chief "comet-catcher;" but he catches them not for himself, but for the sun. Yet if comets do come originally from without the borders of the planetary system, it does not, by any means, follow that they were wanderers at large in space before they yielded to the overmastering attraction of the sun. Investigation of the known cometary orbits, combined with theoretical considerations, has led some astronomers to the conclusion that as the sun travels onward through space he "picks up *en route*" cometary masses which, without belonging strictly to his empire, are borne along in the same vast "cosmical current" that carries the solar system.

But while no intelligent person any longer thinks that the appearance of a great comet is a token from the heavenly powers of the approaching death of a mighty ruler, or the outbreak of a devastating war, or the infliction of a terrible plague upon wicked mankind, science itself has discovered mysteries about comets which are not less fascinating because they are more intellectual than the irrational fancies that they have displaced. To bring the subject properly before the mind, let us see what the principal phenomena connected with a comet are.

At the present day comets are ordinarily "picked up" with the telescope or the photographic plate before anyone except their discoverer is aware of their existence, and usually they remain so insignificant in appearance that only astronomers ever see them. Yet so great is the prestige of the word "comet" that the discovery of one of these inconspicuous wanderers, and its subsequent movements, become items of the day's news which everybody reads with the feeling, perhaps, that at least he knows what is going on in the universe even if he doesn't understand it. But a truly great comet presents quite a different proposition. It, too, is apt to be detected coming out of the depths of space before the world at large can get a glimpse of it, but as it approaches the sun its aspect undergoes a marvelous change. Agitated apparently by solar influence, it throws out a long streaming tail of nebulous light, directed away from the sun and looking as if blown out like a pennon by a powerful wind. Whatever may be the position of the comet with regard to the sun, as it circles round him it continually keeps its tail on the off side. This, as we shall soon see, is a fact of capital importance in relation to the probable nature of comets' tails. Almost at the same time that the formation of the tail is observed a remarkable change takes place in the comet's head, which, by the way, is invariably and not merely occasionally its most important part. On approaching the sun the head usually contracts. Coincidently with this contraction a nucleus generally makes its appearance. This is a bright, starlike point in the head, and it probably represents the totality of solid matter that the comet possesses. But it is regarded as extremely unlikely that even the nucleus consists of a uniformly solid mass. If it were such, comets would be far more formidable visitors when they pass near the planets than they have been found to be. The diameter of the nucleus may vary from a few hundred up to several thousand miles; the heads, on the average, are from twenty-five thousand to one hundred thousand miles in diameter, although a few have greatly exceeded these dimensions; that of the comet of 1811, one of the most stupendous ever seen, was a million and a quarter miles in diameter! As to the tails, not withstanding their enormous length — some have been more than a hundred million miles long — there is reason to believe that they are of extreme tenuity, "as rare as vacuum." The smallest stars have been seen shining through their most brilliant portions with undiminished luster.

After the nucleus has been formed it begins to throw out bright jets directed toward the sun. A stream, and sometimes several streams, of light also project sunward from the nucleus, occasionally appearing like a stunted tail directed oppositely to the real tail. Symmetrical envelopes which, seen in section, appear as half circles or parabolas, rise sunward from the nucleus, forming a concentric series. The ends of these stream backward into the tail, to which they seem to supply material. Ordinarily the formation of these ejections and envelopes is attended by intense agitation of the nucleus, which twists and turns, swinging and gyrating with an appearance of the greatest violence. Sometimes the nucleus is seen to break up into several parts. The entire heads of some comets have been split asunder in passing close around the sun; The comet of 1882 retreated into space after its perihelion passage with *five heads* instead of the one that it had originally, and each of these heads had its own tail!

The possession of the spectroscope has enabled astronomers during later years to study the chemical composition of comets by analyzing their light. At first the only substances thus discovered in them were hydro-carbon compounds, due evidently to the gaseous envelopes in which some combination of hydrogen with carbon existed. Behind this gaseous spectrum was found a faint continuous spectrum ascribed to the nucleus, which apparently both reflects the sunlight and gives forth the light of a glowing solid or liquid. Subsequently sodium and iron lines were found in cometary spectra. The presence of iron would seem to indicate that some of these bodies may be much more massive than observations on their attractive effects have indicated. In some recent comets, such as Morehouse's, in 1908, several lines have been found, the origin of which is unknown.

Without going back of the nineteenth century we may find records of some of the most extraordinary comets that man has ever looked upon. In 1811, still spoken of as "the year of the comet," because of the wonderful vintage ascribed to the skyey visitor, a comet shaped like a gigantic sword amazed the whole world, and, as it remained visible for seventeen months, was regarded by superstitious persons as a symbol of the fearful happenings of Napoleon's Russian campaign. This comet, the extraordinary size of whose head, greatly exceeding that of the sun itself, has already been mentioned, was also remarkable for exhibiting so great a brilliancy without approaching even to the earth's distance from the sun. But there was once a comet (and only once — in the year 1729) which never got nearer to the sun than four times the distance of the earth and yet appeared as a formidable object in the sky. As Professor Young has remarked, "it must have been an enormous comet to be visible from such a distance." And we are to remember that there were no great telescopes in the year 1729. That comet affects the imagination like a phantom of space peering into the solar system, displaying its enormous train afar off (which, if it had approached as near as other comets, would probably have become *the* celestial wonder of all human memory), and then turning away and vanishing in the depths of immensity.

In 1843 a comet appeared which was so brilliant that it could be seen in broad day close beside the sun! This was the first authenticated instance of that kind, but the occurrence was to be repeated, as we shall see in a moment, less than forty years later.

The splendid comet of 1858, usually called Donati's, is remembered by many persons yet living. It was, perhaps, both as seen by the naked eye and with the telescope, the most beautiful comet of which we have any record. It too marked a rich vintage year, still remembered in the vineyards of France, where there is a popular belief that a great comet ripens the grape and imparts to the wine a flavor not attainable by the mere skill of the cultivator. There are "comet wines," carefully treasured in certain cellars, and brought forth only when their owner wishes to treat his guests to a sip from paradise.

The year 1861 saw another very remarkable comet, of an aspect strangely vast and diffuse, which is believed to have swept the earth with its immense tail when it passed between us and the sun on the night of June 30th, an event which produced no other known effect than the appearance of an unwonted amount of scattered light in the sky.

The next very notable comet was the "Great Southern Comet" of 1880, which was not seen from the northern hemisphere. It mimicked the aspect of the famous comet of 1843, and to the great surprise of astronomers appeared to be traveling in the same path. This proved to be the rising of the curtain for an astronomical sensation unparalleled in its kind; for two years later another brilliant comet appeared, first in the southern hemisphere, *and it too followed the same track.* The startling suggestion was now made that this comet was identical with those of 1843 and 1880, its return having been hastened by the resistance experienced in passing twice through the coronal envelope, and there were some who thought that it would now swing swiftly round and then plunge straight into the sun, with consequences that might be disastrous to us on account of the "flash of heat" that would be produced by the impact. Nervous people were frightened, but observation soon proved that the danger was imaginary, for although the comet almost grazed the sun, and must have rushed through two or three million miles of the coronal region, no retardation of its immense velocity was perceptible, and it finally passed away in a damaged condition, as before remarked, and has never since appeared.

Then the probable truth was perceived — *viz.,* that the three comets (1843, 1880, and 1882) were not one identical body, but three separate ones all traveling in the same orbit. It was found, too, that a comet seen in 1668 bore similar insignia of relationship. The natural inference was that these four bodies had once formed a single mass which had been split apart by the disruptive action of the sun. Strength was lent to this hypothesis by the fact that the comet of 1882 was apparently torn asunder during its perihelion passage, retreating into space in a dissevered state. But Prof. George Forbes has a theory that the splitting of the original cometary mass was effected by an unknown planet, probably greater than Jupiter, situated at a hundred times the earth's distance from the sun, and revolving in a period of a thousand years. He supposes that the original comet was not that of 1668, but one seen in 1556, which has since been "missing," and that its disruption occurred from an encounter with the supposititious planet about the year 1700. Truly from every point of view comets are the most extraordinary of adventurers!

The comet of 1882 was likewise remarkable for being visible, like its predecessor of 1843, in full daylight in close proximity to the sun. The story of its detection when almost in contact with the solar disk is dramatic. It had been discovered in the southern hemisphere only a couple of weeks before its perihelion, which occurred on September 17th, and on the forenoon of that day it was seen by Doctor Common in England, and by Doctor Elkin and Mr. Finlay at the Cape of Good Hope, almost touching the sun. It looked like a dazzling white bird with outspread wings. The southern observers watched it go *right into the sun,* when it instantly disappeared. What had happened was that the comet in passing its perihelion point had swung exactly between the earth and the sun. On the following morning it was seen from all parts of the world close by the sun on the opposite side, and it remained thus visible for three days, gradually receding from the solar disk. It then became visible for northern observers in the morning sky before sunrise, brandishing a portentous sword-shaped tail which, if it had been in the evening sky, would have excited the wonder of hundreds of millions, but situated where it was, comparatively few ever saw it.

The application of photography to the study of comets has revealed many curious details which might otherwise have escaped detection, or at best have remained subject to doubt. It has in particular shown not only the precise form of the tails, but the remarkable vicissitudes that they undergo. Professor Barnard's photographs of Brooks' comet in 1893 suggested, by the extraordinary changes in the form of the tail which they revealed, that the comet was encountering a series of obstructions in space which bent and twisted its tail into fantastic shapes. The reader will observe the strange form into which the tail was thrown on the night of October 21st. A cloud of meteors through which the comet was passing might have produced such deformations of its tail. In the photograph of Daniels' comet of 1907, a curious striping of the tail will be noticed. The short bright streaks seen in the photograph, it may be explained, are the images of stars which are drawn out into lines in consequence of the fact that the photographic telescope was adjusted to follow the motion of the comet while the stars remained at rest.

But the adventures of comets are not confined to possible encounters with unknown obstacles. We have referred to the fact that the great planets, and especially Jupiter, frequently interfere with the motions of comets. This interference is not limited to the original alteration of their orbits from possible parabolas to ellipses, but is sometimes exercised again and again, turning the bewildered comets into elliptical paths of all degrees of eccentricity. A famous example of this kind of planetary horse-play is furnished by the story of Lexell's missing comet. This comet was first seen in 1770. Investigation showed that it was moving in an orbit which should bring it back to perihelion every five and a half years; yet it had never been seen before and, although often searched for, has never been seen since. Laplace and Leverrier proved mathematically that in 1767 it had approached so close to Jupiter as to be involved among the orbits of his satellites. What its track had been before is not known, but on that occasion the giant planet seized the interloper, threw it into a short elliptic orbit and sent it, like an arrested vagrant, to receive sentence at the bar of the sun. On this journey it passed within less than 1,500,000 miles of the earth. The form of orbit which Jupiter had impressed required, as we have said, its return in about five and a half years; but soon after 1770 it had the misfortune a second time to encounter Jupiter at close range, and he, as if dissatisfied with the leniency of the sun, or indignant at the stranger's familiarity, seized the comet and hurled it out of the system, or at any rate so far away that it has never since been able to rejoin the family circle that basks in the immediate rays of the solar hearth. Nor is this the only instance in which Jupiter has dealt summarily with small comets that have approached him with too little deference.

The function which Jupiter so conspicuously fulfills as master of the hounds to the sun is worth considering a little more in detail. To change the figure, imagine the sun in its voyage through space to be like a majestic battleship surrounded by its scouts. Small vessels (the comets, as they are overhauled by the squadron, are taken in charge by the scouts, with Jupiter for their chief, and are forced to accompany the fleet, but not all are impressed. If a strange comet undertakes to run across Jupiter's bows the latter brings it to, and makes prize of it by throwing it into a relatively small ellipse with the sun for its focus. Thenceforth, unless, as happened to the unhappy comet of Lexell, it encounters Jupiter again in such a way as to be diverted by him into a more distant orbit, it can never get away. About thirty comets are now known to have thus been captured by the great planet, and they are called "Jupiter's Comet Family." But, on the other hand, if a wandering comet crosses the wake of the chief planetary scout the latter simply drives it away by accelerating its motion and compels it to steer off into open space. The transformation of comets into meteors will be considered in the next chapter, but here, in passing, mention may be made of the strange fate of one member of Jupiter's family, Biela's comet, which, having become over bold in its advances to its captor, was, after a few revolutions in is impressed orbit, torn to pieces and turned into a flock of meteors.

And now let us return to the mystery of comets' tails. That we are fully justified in speaking of the tails of comets as mysterious is proved by the declaration of Sir John Herschel, who averred, in so many words, that "there is some profound secret and mystery of nature concerned in this phenomenon," and this profound secret and mystery has not yet been altogether cleared up. Nevertheless, the all-explaining hypothesis of Arrhenius offers us once more a certain amount of aid. Comets' tails, Arrhenius assures us, are but another result of the pressure of light. The reader will recall the applications of this theory to the Zodiacal Light and the Aurora. In the form in which we now have to deal with it, the supposition is made that as a comet approaches the sun eruptions of vapor, due to the solar heat, occur in its nucleus. These are naturally most active on the side which is directly exposed to the sun, whence the appearance of the immense glowing envelopes that surround the nucleus on the sunward side. Among the particles of hydro-carbon, and perhaps solid carbon in the state of fine dust, which are thus set free there will be many whose size is within the critical limit which enables the light-waves from the sun to drive them away. Clouds of such particles, then, will stream off behind the advancing comet, producing the appearance of a tail. This accounts for the fact that the tails of comets are always directed away from the sun, and it also explains the varying forms of the tails and the extraordinary changes that they undergo. The speed of the particles driven before the light-waves must depend upon their size and weight, the lightest of a given size traveling the most swiftly. By accretion certain particles might grow, thus losing velocity and producing the appearance of bunches in the tail, such as have been observed. The hypothesis also falls in with the researches of Bredichin, who has divided the tails of comets into three principal classes — *viz.:* (1) Those which appear as long, straight rays; (2) Those which have the form of curved plumes or scimitars; (3) Those which are short, brushy, and curved sharply backward along the comet's path. In the first type he calculates the repulsive force at from twelve to fifteen times the force of gravity; in the second at from two to four times; and in the third at about one and a half times. The straight tails he ascribes to hydrogen because the hydrogen atom is the lightest known; the sword-shaped tails to hydro-carbons; and the stumpy tails to vaporized iron. It will be seen that, if the force driving off the tails is that which Arrhenius assumes it to be, the forms of those appendages would accord with those that Bredichin's theory calls for. At the same time we have an explanation of the multiple tails with which some comets have adorned themselves. The comet of 1744, for instance, had at one time no less than seven tails spread in a wide curved brush behind it. Donati's comet of 1858 also had at least two tails, the principal one sword-shaped and the other long, narrow, and as straight as a rule. According to Bredichin, the straight tail must have been composed of hydrogen, and the other of some form of hydro-carbon whose atoms are heavier than those of hydrogen, and, consequently, when swept away by the storm of light-waves, followed a curvature depending upon the resultant of the forces operating upon them. The seven tails of the comet of 1744 presented a kind of diagram graphically exhibiting its complex composition, and, if we knew a little more about the constituents of a comet, we might be able to say from the amount of curvature of the different tails just what were the seven substances of which that comet consisted.

If these theories seem to the reader fantastic, at any rate they are no more fantastic than the phenomena that they seek to explain.

Meteors, Fire-Balls, and Meteorites

One of the most terrorizing spectacles with which the heavens have ever caused the hearts of men to quake occurred on the night of November 13, 1833. On that night North America, which faced the storm, was under a continual rain of fire from about ten o'clock in the evening until daybreak.

The fragments of a comet had struck the earth.

But the meaning of what had happened was not discovered until long afterward. To the astronomers who, with astonishment not less than that of other people, watched the wonderful scene, it was an unparalleled "shower of meteors." They did not then suspect that those meteors had once formed the head of a comet. Light dawned when, a year later, Prof. Denison Olmsted, of Yale College, demonstrated that the meteors had all moved in parallel orbits around the sun, and that these orbits intersected that of the earth at the point where our planet happened to be on the memorable night of November 13th. Professor Olmsted even went so far as to suggest that the cloud of meteors that had encountered the earth might form a diffuse comet; but full recognition of the fact that they were cometary débris came later, as the result of further investigation. The key to the secret was plainly displayed in the spectacle itself, and was noticed without being understood by thousands of the terror-stricken beholders. It was *an umbrella of fire* that had opened overhead and covered the heavens; in other words, the meteors all radiated from a particular point in the constellation Leo, and, being countless as the snowflakes in a winter tempest, they ribbed the sky with fiery streaks. Professor Olmsted showed that the radiation of the meteors from a fixed point was an effect of perspective, and in itself a proof that they were moving in parallel paths when they encountered the earth. The fact was noted that there had been a similar, but incomparably less brilliant, display of meteors on the same day of November, 1832, and it was rightly concluded that these had belonged to the same stream, although the true relationship of the phenomena was not immediately apprehended. Olmsted ascribed to the meteors a revolution about the sun once in every six months, bringing them to the intersection of their orbit with that of the earth every November 13th; but later investigators found that the real period was about thirty-three and one-quarter years, so that the great displays were due three times in a century, and their return was confidently predicted for the year 1866. The appearance of the meteors in 1832, a year before the great display, was ascribed to the great length of the stream which they formed in space — so great that they required more than two years to cross the earth's orbit. In 1832 the earth had encountered a relatively rare part of the stream, but in 1833, on returning to the crossing-place, it found there the richest part of the stream pouring across its orbit. This explanation also proved to be correct, and the predicted return in 1866 was duly witnessed, although the display was much less brilliant than in 1833. It was followed by another in 1867.

In the meantime Olmsted's idea of a cometary relationship of the meteors was demonstrated to be correct by the researches of Schiaparelli and others, who showed that not only the November meteors, but those of August, which are seen more or less abundantly every year, traveled in the tracks of well-known comets, and had undoubtedly an identical origin with those comets. In other words the comets and the meteor-swarms were both remnants of original masses which had probably been split up by the action of the sun, or of some planet to which they had made close approaches. The annual periodicity of the August meteors was ascribed to the fact that the separation had taken place so long ago that the meteors had become distributed all around the orbit, in consequence of which the earth encountered some of them every year when it arrived at the crossing-point. Then Leverrier showed that the original comet associated with the November meteors was probably brought into the system by the influence of the planet Uranus in the year 126 of the Christian era. Afterward Alexander Herschel identified the tracks of no less than seventy-six meteor-swarms (most of them inconspicuous) with those of comets. The still more recent researches of Mr. W.F. Denning make it probable that there are no meteors which do not belong to a flock or system probably formed by the disintegration of a cometary mass; even the apparently sporadic ones which shoot across the sky, "lost souls in the night," being members of flocks which have become so widely scattered that the earth sometimes takes weeks to pass through the region of space where their paths lie.

The November meteors should have exhibited another pair of spectacles in 1899 and 1900, and their failure to do so caused at first much disappointment, until it was made plain that a good reason existed for their absence. It was found that after their last appearance, in 1867, they had been disturbed in their movements by the planets Jupiter and Saturn, whose attractions had so shifted the position of their orbit that it no longer intersected that of the earth, as it did before. Whether another planetary interference will sometime bring the principal mass of the November meteors back to the former point of intersection with the earth's orbit is a question for the future to decide. It would seem that there may be several parallel streams of the November meteors, and that some of them, like those of August, are distributed entirely around the orbit, so that every mid-November we see a few of them.

We come now to a very remarkable example of the disintegration of a comet and the formation of a meteor-stream. In 1826 Biela, of Josephstadt, Austria, discovered a comet to which his name was given. Calculation showed that it had an orbital period of about six and a half years, belonging to Jupiter's "family." On one of its returns, in 1846, it astonished its watchers by suddenly splitting in two. The two comets thus formed out of one separated to a distance of about one hundred and sixty thousand miles, and then raced side by side, sometimes with a curious ligature connecting them, like Siamese twins, until they disappeared together in interplanetary space. In 1852 they came back, still nearly side by side, but now the distance between them had increased to a million and a quarter of miles. After that, at every recurrence of their period, astronomers looked for them in vain, until 1872, when an amazing thing happened. On the night of November 28th, when the earth was crossing the plane of the orbit of the missing comet, a brilliant shower of meteors burst from the northern sky, traveling nearly in the track which the comet should have pursued. The astronomers were electrified. Klinkerfues, of Göttingen, telegraphed to Pogson, of Madras: *"Biela touched earth; search near Theta Centauri."* Pogson searched in the place indicated and saw a cometary mass retreating into the southern heavens, where it was soon swallowed from sight!

Since then the Biela meteors have been among the recognized periodic spectacles of the sky, and few if any doubt that they represent a portion of the missing comet whose disintegration began with the separation into two parts in 1846. The comet itself has never since been seen. The first display of these meteors, sometimes called the "Andromedes," because they radiate from the constellation Andromeda, was remarkable for the great brilliancy of many of the fire-balls that shot among the shower of smaller sparks, some of which were described as equaling the full moon in size. None of them is known to have reached the earth, but during the display of the same meteors in 1885 a meteoric mass fell at Mazapil in Northern Mexico (it is now in the Museum at Vienna), which many have thought may actually be a piece of the original comet of Biela. This brings us to the second branch of our subject.

More rare than meteors or falling stars, and more startling, except that they never appear in showers, are the huge balls of fire which occasionally dart through the sky, lighting up the landscapes beneath with their glare, leaving trains of sparks behind them, often producing peals of thunder when they explode, and in many cases falling upon the earth and burying themselves from a few inches to several feet in the soil, from which, more than once, they have been picked up while yet hot and fuming. These balls are sometimes called bolides. They are not really round in shape, although they often look so while traversing the sky, but their forms are fragmentary, and occasionally fantastic. It has been supposed that their origin is different from that of the true meteors; it has even been conjectured that they may have originated from the giant volcanoes of the moon or have been shot out from the sun during some of the tremendous explosions that accompany the formation of eruptive prominences. By the same reasoning some of them might be supposed to have come from some distant star. Others have conjectured that they are wanderers in space, of unknown origin, which the earth encounters as it journeys on, and Lord Kelvin made a suggestion which has become classic because of its imaginative reach — *viz.,* that the first germs of life may have been brought to the earth by one of these bodies, "a fragment of an exploded world."

It is a singular fact that astronomers and scientific men in general were among the last to admit the possibility of solid masses falling from the sky. The people had believed in the reality of such phenomena from the earliest times, but the savants shook their heads and talked of superstition. This was the less surprising because no scientifically authenticated instance of such an occurrence was known, and the stones popularly believed to have fallen from the sky had become the objects of worship or superstitious reverence, a fact not calculated to recommend them to scientific credence. The celebrated "black stone" suspended in the Kaaba at Mecca is one of these reputed gifts from heaven; the "Palladium" of ancient Troy was another; and a stone which fell near En-sisheim, in Germany, was placed in a church as an object to be religiously venerated. Many legends of falling stones existed in antiquity, some of them curiously transfigured by the imagination, like the "Lion of the Peloponnesus," which was said to have sprung down from the sky upon the Isthmus of Corinth. But near the beginning of the nineteenth century, in 1803, a veritable shower of falling stones occurred at L'Aigle, in Northern France, and this time astronomers took note of the phenomenon and scientifically investigated it.

Thousands of the strange projectiles came from the sky on this occasion, and were scattered over a wide area of country, and some buildings were hit. Four years later another shower of stones occurred at Weston, Conn., numbering thousands of individuals. The local alarm created in both cases was great, as well it might be, for what could be more intimidating than to find the blue vault of heaven suddenly hurling solid missiles at the homes of men? After these occurrences it was impossible for the most skeptical to doubt any longer, and the regular study of "aerolites," or "meteorites," began.

One of the first things recognized was the fact that fire-balls are solid meteorites in flight, and not gaseous exhalations in the air, as some had assumed. They burn in the air during their flight, and sometimes, perhaps, are entirely consumed before reaching the ground. Their velocity before entering the earth's atmosphere is equal to that of the planets in their orbits — *viz.,* from twenty to thirty miles per second — a fact which proves that the sun is the seat of the central force governing them. Their burning in the air is not difficult to explain; it is the heat of friction which so quickly brings them to incandescence. Calculation shows that a body moving through the air at a velocity of about a mile per second will be brought, superficially, to the temperature of "red heat" by friction with the atmosphere. If its velocity is twenty miles per second the temperature will become thousands of degrees. This is the state of affairs with a meteorite rushing into the earth's atmosphere; its surface is liquefied within a few seconds after the friction begins to act, and the melted and vaporized portion of its mass is swept backward, forming the train of sparks that follows every great fire-ball. However, there is one phenomenon connected with the trains of meteorites which has never been satisfactorily explained: they often persist for long periods of time, drifting and turning with the wind, but not ceasing to glow with a phosphorescent luminosity. The question is, Whence comes this light? It must be light without heat, since the fine dust or vapor of which the train can only consist would not retain sufficient heat to render it luminous for so long a time. An extremely remarkable incident of this kind occurred on February 22, 1909, when an immense fire-ball that passed over southern England left a train that remained visible during two hours, assuming many curious shapes as it was drifted about by currents in the air.

But notwithstanding the enormous velocity with which meteorites enter the air they are soon slowed down to comparatively moderate speed, so that when they disappear they are usually traveling not faster than a mile a second. The courses of many have been traced by observers situated along their track at various points, and thus a knowledge has been obtained of their height above the ground during their flight and of the length of their visible courses. They generally appear at an elevation of eighty or a hundred miles, and are seldom visible after having descended to within five miles of the ground, unless the observer happens to be near the striking-point, when he may actually witness the fall. Frequently they burst while high in the air and their fragments are scattered like shrapnel over the surface of the ground, sometimes covering an area of several square miles, but of course not thickly; different fragments of the same meteorite may reach the ground at points several miles apart. The observed length of their courses in the atmosphere varies from fifty to five hundred miles. If they continued a long time in flight after entering the air, even the largest of them would probably be consumed to the last scrap, but their fiery career is so short on account of their great speed that the heat does not have time to penetrate very deeply, and some that have been picked up immediately after their fall have been found cold as ice within. Their size after reaching the ground is variable within wide limits; some are known which weigh several tons, but the great majority weigh only a few pounds and many only a few ounces.

Meteorites are of two kinds: *stony* meteorites and *iron* meteorites. The former outnumber the latter twenty to one; but many stone meteorites contain grains of iron. Nickel is commonly found in iron meteorites, so that it might be said that that redoubtable alloy nickel-steel is of cosmical invention. Some twenty-five chemical elements have been found in meteorites, including carbon and the "sun-metal," helium. The presence of the latter is certainly highly suggestive in connection with the question of the origin of meteorites. The iron meteorites, besides metallic iron and nickel, of which they are almost entirely composed, contain hydrogen, helium, and carbonic oxide, and about the only imaginable way in which these gases could have become absorbed in the iron would be through the immersion of the latter while in a molten or vaporized state in a hot and dense atmosphere composed of them, a condition which we know to exist only in the envelopes of the sun and the stars.

The existence of carbon in the Canyon Diablo iron meteorites is attended by a circumstance of the most singular character — a very "fairy tale of science." In some cases *the carbon has become diamond!* These meteoric diamonds are very small; nevertheless, they are true diamonds, resembling in many ways the little black gems produced by Moissan's method with the aid of the electric furnace. The fact that they are found embedded in these iron meteorites is another argument in favor of the hypothesis of the solar or stellar origin of the latter. To appreciate this it is necessary to recall the way in which Moissan made his diamonds. It was by a combination of the effects of great heat, great pressure, and sudden or rapid superficial cooling on a mass of iron containing carbon. When he finally broke open his iron he found it a pudding stuffed with miniature black diamonds. When a fragment of the Canyon Diablo meteoric iron was polished in Philadelphia over fifteen years ago it cut the emery-wheel to pieces, and examination showed that the damage had been effected by microscopic diamonds peppered through the mass. How were those diamonds formed? If the sun or Sirius was the laboratory that prepared them, we can get a glimpse at the process of their formation. There is plenty of heat, plenty of pressure, and an abundance of vaporized iron in the sun and the stars. When a great solar eruption takes place, masses of iron which have absorbed carbon may be shot out with a velocity which forbids their return. Plunged into the frightful cold of space, their surfaces are quickly cooled, as Moissan cooled his prepared iron by throwing it into water, and thus the requisite stress is set up within, and, as the iron solidifies, the included carbon crystallizes into diamonds. Whether this explanation has a germ of truth in it or not, at any rate it is evident that iron meteorites were not created in the form in which they come to us; they must once have been parts of immeasurably more massive bodies than themselves.

The fall of meteorites offers an appreciable, though numerically insignificant, peril to the inhabitants of the earth. Historical records show perhaps three or four instances of people being killed by these bodies. But for the protection afforded by the atmosphere, which acts as a very effective shield, the danger would doubtless be very much greater. In the absence of an atmosphere not only would more meteorites reach the ground, but their striking force would be incomparably greater, since, as we have seen, the larger part of their original velocity is destroyed by the resistance of the air. A meteorite weighing many tons and striking the earth with a velocity of twenty or thirty miles per second, would probably cause frightful havoc.

It is a singular fact that recent investigations seem to have proved that an event of this kind actually happened in North America — perhaps not longer than a thousand or two thousand years ago. The scene of the supposed catastrophe is in northern central Arizona, at Coon Butte, where there is a nearly circular crater in the middle of a circular elevation or small mountain. The crater is somewhat over four thousand feet in diameter, and the surrounding rim, formed of upturned strata and ejected rock fragments, rises at its highest point one hundred and sixty feet above the plain. The crater is about six hundred feet in depth — that is, from the rim to the visible floor or bottom of the crater. There is no evidence that volcanic action has ever taken place in the immediate neighborhood of Coon Butte. The rock in which the crater has been made is composed of horizontal sandstone and limestone strata. Between three hundred and four hundred million tons of rock fragments have been detached, and a large portion hurled by some cause out of the crater. These fragments lie concentrically distributed around the crater, and in large measure form the elevation known as Coon Butte. The region has been famous for nearly twenty years on account of the masses of meteoric iron found scattered about and known as the "Canyon Diablo" meteorites. It was one of these masses, which consist of nickel-iron containing a small quantity of platinum, and of which in all some ten tons have been recovered for sale to the various collectors through-out the world, that as before mentioned destroyed the grind-ing-tool at Philadelphia through the cutting power of its embedded diamonds. These meteoric irons are scattered about the crater-hill, in concentric distribution, to a maxi-mum distance of about five miles. When the suggestion was first made in 1896 that a monster meteorite might have created by its fall this singular lone crater *in stratified rocks,* it was greeted with incredulous smiles; but since then the matter has assumed a different aspect. The Standard Iron Company, formed by Messrs. D.M. Barringer, B.C. Tilghman, E.J. Ben-nitt, and S.J. Holsinger, having become, in 1903, the owner of this freak of nature, sunk shafts and bored holes to a great depth in the interior of the crater, and also trenched the slopes of the mountain, and the result of their investigations has proved that the meteoric hypothesis of origin is correct. (See the papers published in the *Proceedings of the Academy of Natural Sciences of Philadelphia,* December, 1905, wherein it is proved that the United States Geological Survey was wrong in believing this crater to have been due to a steam explosion. Since that date there has been discovered a great amount of additional confirmatory proof). Material of unmistakably meteoric origin was found by means of the drills, mixed with crushed rock, to a depth of six hundred to seven hundred feet below the floor of the crater, and a great deal of it has been found admixed with the ejected rock fragments on the outer slopes of the mountain, absolutely proving synchronism between the two events, the formation of this great crater and the falling of the meteoric iron out of the sky. The drill located in the bottom of the crater was sent, in a number of cases, much deeper (over one thousand feet) into unaltered horizontal red sandstone strata, but no meteoric material was found below this depth (seven hundred feet, or between eleven and twelve hundred feet below the level of the sur-rounding plain), which has been assumed as being about the limit of penetration. It is not possible to sink a shaft at present, owing to the water which has drained into the crater, and which forms, with the finely pulverized sandstone, a very troublesome quicksand encountered at about two hundred feet below the visible floor of the crater. As soon as this water is removed by pumping it will be easy to explore the depths of the crater by means of shafts and drifts. The rock strata (sandstone and limestone) of which the walls consist present every appearance of having been violently upturned by a huge body penetrating the earth like a cannon-ball. The general aspect of the crater strikingly resembles the impression made by a steel projectile shot into an armor-plate. Mr. Tilghman has estimated that a meteorite about five hundred feet in diameter and moving with a velocity of about five miles per second would have made just such a perforation upon strik-ing rocks of the character of those found at this place. There was some fusion of the colliding masses, and the heat pro-duced some steam from the small amount of water in the rocks. As a result there has been found at depth a considerable amount of fused quartz (original sandstone), and with it innumerable particles or sparks of fused nickel-iron (original meteorite). A projectile of that size penetrating eleven to twelve hundred feet into the rocky shell of the globe must have produced a shock which was perceptible several hundred miles away.

The great velocity ascribed to the supposed meteorite at the moment of striking could be accounted for by the fact that it probably plunged nearly vertically downward, for it formed a circular crater in the rocky crust of the earth. In that case it would have been less retarded by the resistance of the atmosphere than are meteorites which enter the air at a lower angle and shoot ahead hundreds of miles until friction has nearly destroyed their original motion when they drop upon the earth. Some meteoric masses of great size, such as Peary's iron meteorite found at Cape York, Greenland, and the almost equally large mass discovered at Bacubirito, Mex-ico, appear to have penetrated but slightly on striking the earth. This may be explained by supposing that they pursued a long, horizontal course through the air before falling. The result would be that, their original velocity having been practically destroyed, they would drop to the ground with a velocity nearly corresponding to that which gravity would impart within the perpendicular distance of their final fall. A six-hundred-and-sixty-pound meteorite, which fell at Knyahinya, Hungary, striking at an angle of 27° from the vertical, penetrated the ground to a depth of eleven feet.

It has been remarked that the Coon Butte meteorite may have fallen not longer ago than a few thousand years. This is based upon the fact that the geological indications favor the supposition that the event did not occur more than five thousand years ago, while on the other hand the rings of growth in the cedar trees growing on the slopes of the crater show that they have existed there about seven hundred years. Prof. William H. Pickering has recently correlated this with an ancient chronicle which states that at Cairo, Egypt, in the year 1029, "many stars passed with a great noise." He remarks that Cairo is about 100°, by great circle, from Coon Butte, so that if the meteorite that made the crater was a member of a flock of similar bodies which encountered the earth moving in parallel lines, some of them might have traversed the sky tangent to the earth's surface at Cairo. That the spectacle spoken of in the chronicle was caused by meteorites he deems exceedingly probable because of what is said about "a great noise;" meteorites are the only celestial phenomena attended with perceptible sounds. Professor Pickering conjectures that this supposed flock of great meteorites may have formed the nucleus of a comet which struck the earth, and he finds confirmation of the idea in the fact that out of the ten largest meteorites known, no less than seven were found within nine hundred miles of Coon Butte. It would be interesting if we could trace back the history of that comet, and find out what malicious planet caught it up in its innocent wanderings and hurled it with so true an aim at the earth! This remarkable crater is one of the most interesting places in the world, for there is absolutely no record of such a mass, possibly an iron-headed comet, from outer space having come into collision with our earth. The results of the future exploration of the depths of the crater will be awaited with much interest.

The Wrecking of the Moon

There are sympathetic moods under whose influence one gazes with a certain poignant tenderness at the worn face of the moon; that little "fossil world" (the child of our mother earth, too) bears such terrible scars of its brief convulsive life that a sense of pity is awakened by the sight. The moon is the wonder-land of the telescope. Those towering mountains, whose "proud aspiring peaks" cast silhouettes of shadow that seem drawn with India-ink; those vast plains, enchained with gentle winding hills and bordered with giant ranges; those oval "oceans," where one looks expectant for the flash of wind-whipped waves; those enchanting "bays" and recesses at the seaward feet of the Alps; those broad straits passing between guardian heights incomparably mightier than Gibraltar; those locketlike valleys as secluded among their mountains as the Vale of Cashmere; those colossal craters that make us smile at the pretensions of Vesuvius, Etna, and Cotopaxi; those strange white ways which pass with the unconcern of Roman roads across mountain, gorge, and valley – all these give the beholder an irresistible impression that it is truly a world into which he is looking, a world akin to ours, and yet no more like our world than Pompeii is like Naples. Its air, its waters, its clouds, its life are gone, and only a skeleton remains – a mute but eloquent witness to a cosmical tragedy without parallel in the range of human knowledge.

One cannot but regret that the moon, if it ever was the seat of intelligent life, has not remained so until our time. Think what the consequences would have been if this other world at our very door had been found to be both habitable and inhabited! We talk rather airily of communicating with Mars by signals; but Mars never approaches nearer than 35,000,000 miles, while the moon when nearest is only a little more than 220,000 miles away. Given an effective magnifying power of five thousand diameters, which will perhaps be possible at the mountain observatories as telescopes improve, and we should be able to bring the moon within an apparent distance of about forty miles, while the corresponding distance for Mars would be more than seven thousand miles. But even with existing telescopic powers we can see details on the moon no larger than some artificial constructions on the earth. St Peter's at Rome, with the Vatican palace and the great piazza, if existing on the moon, would unquestionably be recognizable as something else than a freak of nature. Large cities, with their radiating lines of communication, would at once betray their real character. Cultivated tracts, and the changes produced by the interference of intelligent beings, would be clearly recognizable. The electric illumination of a large town at night would probably be markedly visible. Gleams of reflected sunlight would come to us from the surfaces of the lakes and oceans, and a huge "liner" traversing a lunar sea could probably be followed by its trail of smoke. As to communications by "wireless" signals, which certain enthusiasts have thought of in connection with Mars, in the case of the moon they should be a relatively simple matter, and the feat might actually be accomplished. Think what a literature would grow up about the moon if it were a living world! Its very differences from the earth would only accentuate its interest for us. Night and day on the moon are each two weeks in length; how interesting it would be to watch the manner in which the lunarians dealt with such a situation as that. Lunar and terrestrial history would keep step with each other, and we should record them both. Truly one might well wish to have a neighbor world to study; one would feel so much the less alone in space.

It is not impossible that the moon did at one time have inhabitants of some kind. But, if so, they vanished with the disappearance of its atmosphere and seas, or with the advent of its cataclysmic age. At the best, its career as a living world must have been brief. If the water and air were gradually absorbed, as some have conjectured, by its cooling interior rocks, its surface might, nevertheless, have retained them for long ages; but if, as others think, their disappearance was due to the escape of their gaseous molecules in consequence of the inability of the relatively small lunar gravitation to retain them, then the final catastrophe must have been as swift as it was inevitable. Accepting Darwin's hypothesis, that the moon was separated from the earth by tidal action while both were yet plastic or nebulous, we may reasonably conclude that it began its career with a good supply of both water and air, but did not possess sufficient mass to hold them permanently. Yet it may have retained them long enough for life to develop in many forms upon its surface; in fact, there are so many indications that air and water have not always been lacking to the lunar world that we are driven to invent theories to explain both their former presence and their present absence.

But whatever the former condition of the moon may have been, its existing appearance gives it a resistless fascination, and it bears so clearly the story of a vast catastrophe sculptured on its rocky face that the thoughtful observer cannot look upon it without a feeling of awe. The gigantic character of the lunar features impresses the beholder not less than the universality of the play of destructive forces which they attest. Let us make a few comparisons. Take the lunar crater called "Tycho," which is a typical example of its kind. In the telescope Tycho appears as a perfect ring surrounding a circular depression, in the center of which rises a group of mountains. Its superficial resemblance to some terrestrial volcanic craters is very striking. Vesuvius, seen from a point vertically above, would no doubt look something like that (the resemblance would have been greater when the Monte del Cavallo formed a more complete circuit about the crater cone). But compare the dimensions. The remains of the outer crater ring of Vesuvius are perhaps half a mile in diameter, while the active crater itself is only two or three hundred feet across at the most; Tycho has a diameter of fifty-four miles! The group of relatively insignificant peaks in the center of the crater floor of Tycho is far more massive than the entire mountain that we call Vesuvius. The largest known volcanic crater on the earth, Aso San, in Japan, has a diameter of seven miles; it would take *sixty* craters like Aso San to equal Tycho in area! And Tycho, though one of the most perfect, is by no means the largest crater on the moon. Another, called "Theophilus," has a diameter of sixty-four miles, and is eighteen thousand feet deep. There are hundreds from ten to forty miles in diameter, and thousands from one to ten miles. They are so numerous in many places that they break into one another, like the cells of a crushed honeycomb.

The lunar craters differ from those of the earth more fundamentally than in the matter of mere size; *they are not situated on the tops of mountains.* If they were, and if all the proportions were the same, a crater like Tycho might crown a conical peak fifty or one hundred miles high! Instead of being cavities in the summits of mountains, the lunar craters are rather gigantic sink-holes whose bottoms in many cases lie two or three miles below the general surface of the lunar world. Around their rims the rocks are piled up to a height of from a few hundred to two or three thousand feet, with a comparatively gentle inclination, but on the inner side they fall away in gigantic broken precipices which make the dizzy cliffs of the Matterhorn seem but "lover's leaps." Down they drop, ridge below ridge, crag under crag, tottering wall beneath wall, until, in a crater named "Newton," near the south lunar pole, they attain a depth where the rays of the sun never reach. Nothing more frightful than the spectacle which many of these terrible chasms present can be pictured by the imagination. As the lazy lunar day slowly advances, the sunshine, unmitigated by clouds or atmospheric veil of any kind, creeps across their rims and begins to descend the opposite walls. Presently it strikes the ragged crest of a ridge which had lain hidden in such darkness as we never know on the earth, and runs along it like a line of kindling fire. Rocky pinnacles and needles shoot up into the sunlight out of the black depths. Down sinks the line of light, mile after mile, and continually new precipices and cliffs are brought into view, until at last the vast floor is attained and begins to be illuminated. In the meanwhile the sun's rays, darting across the gulf, have touched the summits of the central peaks, twenty or thirty miles from the crater's inmost edge, and they immediately kindle and blaze like huge stars amid the darkness. So profound are some of these awful craters that days pass before the sun has risen high enough above them to chase the last shadows from their depths.

Although several long ranges of mountains resembling those of the earth exist on the moon, the great majority of its elevations assume the crateriform aspect. Sometimes, instead of a crater, we find an immense mountain ring whose form and aspect hardly suggest volcanic action. But everywhere the true craters are in evidence, even on the sea-beds, although they attain their greatest number and size on those parts of the moon — covering sixty percent of its visible surface — which are distinctly mountainous in character and which constitute its most brilliant portions. Broadly speaking, the southwestern half of the moon is the most mountainous and broken, and the northeastern half the least so. Right down through the center, from pole to pole, runs a wonderful line of craters and crateriform valleys of a magnitude stupendous even for the moon. Another similar line follows the western edge. Three or four "seas" are thrust between these mountainous belts. By the effects of "libration" parts of the opposite hemisphere of the moon which is turned away from the earth are from time to time brought into view, and their aspect indicates that that hemisphere resembles in its surface features the one which faces the earth. There are many things about the craters which seem to give some warrant for the hypothesis which has been particularly urged by Mr. G.K. Gilbert, that they were formed by the impact of meteors; but there are also many things which militate against that idea, and, upon the whole, the volcanic theory of their origin is to be preferred.

The enormous size of the lunar volcanoes is not so difficult to account for when we remember how slight is the force of lunar gravity as compared with that of the earth. With equal size and density, bodies on the moon weigh only one-sixth as much as on the earth. Impelled by the same force, a projectile that would go ten miles on the earth would go sixty miles on the moon. A lunar giant thirty-five feet tall would weigh no more than an ordinary son of Adam weighs on his greater planet. To shoot a body from the earth so that it would not drop back again, we should have to start it with a velocity of seven miles per second; a mile and a half per second would serve on the moon. It is by no means difficult to believe, then, that a lunar volcano might form a crater ring eight or ten times broader than the greatest to be found on the earth, especially when we reflect that in addition to the relatively slight force of gravity, the materials of the lunar crust are probably lighter than those of our terrestrial rocks.

For similar reasons it seems not impossible that the theory mentioned in a former chapter — that some of the meteorites that have fallen upon the earth originated from the lunar volcanoes — is well founded. This would apply especially to the stony meteorites, for it is hardly to be supposed that the moon, at least in its superficial parts, contains much iron. It is surely a scene most strange that is thus presented to the mind's eye — that little attendant of the earth's (the moon has only one-fiftieth of the volume, and only one-eightieth of the mass of the earth) firing great stones back at its parent planet! And what can have been the cause of this furious outbreak of volcanic forces on the moon? Evidently it was but a passing stage in its history; it had enjoyed more quiet times before. As it cooled down from the plastic state in which it parted from the earth, it became incrusted after the normal manner of a planet, and then oceans were formed, its atmosphere being sufficiently dense to prevent the water from evaporating and the would-be oceans from disappearing continually in mist. This, if any, must have been the period of life in the lunar world. As we look upon the vestiges of that ancient world buried in the wreck that now covers so much of its surface, it is difficult to restrain the imagination from picturing the scenes which were once presented there; and, in such a case, should the imagination be fettered? We give it free rein in terrestrial life, and it rewards us with some of our greatest intellectual pleasures. The wonderful landscapes of the moon offer it an ideal field with just enough half-hidden suggestions of facts to stimulate its powers.

The great plains of the *Mare Imbrium* and the *Mare Serenitatis* (the "Sea of Showers" and the "Sea of Serenity"), bordered in part by lofty mountain ranges precisely like terrestrial mountains, scalloped along their shores with beautiful bays

curving back into the adjoining highlands, and united by a
great strait passing between the nearly abutting ends of the
"Lunar Apennines" and the "Lunar Caucasus," offer the
elements of a scene of world beauty such as it would be
difficult to match upon our planet. Look at the finely modu-
lated bottom of the ancient sea in Mr. Ritchey's exquisite
photograph of the western part of the *Mare Serenitatis,* where
one seems to see the play of the watery currents heaping the
ocean sands in waving lines, making shallows, bars, and deeps
for the mariner to avoid or seek, and affording a playground
for the creatures of the main. What geologist would not wish
to try his hammer on those rocks with their stony pages of
fossilized history? There is in us an instinct which forbids us
to think that there was never any life there. If we could visit
the moon, there is not among us a person so prosaic and
unimaginative that he would not, the very first thing, begin
to search for traces of its inhabitants. We would look for them
in the deposits on the sea bottoms; we would examine the
shores wherever the configuration seemed favorable for har-
bors and the sites of maritime cities — forgetting that it may
be a little ridiculous to ascribe to the ancient lunarians the
same ideas that have governed the development of our race;
we would search through the valleys and along the seeming
courses of vanished streams; we would explore the mountains,
not the terrible craters, but the pinnacled chains that recall
our own Alps and Rockies; seeking everywhere some vestige
of the transforming presence of intelligent life. Perhaps we
should find such traces, and perhaps, with all our searching,
we should find nothing to suggest that life had ever existed
amid that universal ruin.

Look again at the border of the "Sea of Serenity" — what
a name for such a scene! — and observe how it has been rent
with almost inconceivable violence, the wall of the colossal
crater Posidonius dropping vertically upon the ancient shore
and obliterating it, while its giant neighbor, Le Monnier,
opens a yawning mouth as if to swallow the sea itself. A scene
like this makes one question whether, after all, those may not
be right who have imagined that the so-called sea bottoms
are really vast plains of frozen lava which gushed up in floods
so extensive that even the mighty volcanoes were half

drowned in the fiery sea. This suggestion becomes even stronger when we turn to another of the photographs of Mr. Ritchey's wonderful series, showing a part of the *Mare Tranquilitatis* ("Sea of Tranquility!"). Notice how near the center of the picture the outline of a huge ring with radiating ridges shows through the sea bottom; a fossil volcano submerged in a petrified ocean! This is by no means the only instance in which a buried world shows itself under the great lunar plains. Yet, as the newer craters in the sea itself prove, the volcanic activity survived this other catastrophe, or broke out again subsequently, bringing more ruin to pile upon ruin.

Yet notwithstanding the evidence which we have just been considering in support of the hypothesis that the "seas" are lava floods, Messrs. Loewy and Puiseux, the selenographers of the Paris Observatory, are convinced that these great plains bear characteristic marks of the former presence of immense bodies of water. In that case we should be forced to conclude that the later oceans of the moon lay upon vast sheets of solidified lava; and thus the catastrophe of the lunar world assumes a double aspect, the earliest oceans being swallowed up in molten floods issuing from the interior, while the lands were reduced to chaos by a universal eruption of tremendous volcanoes; and then a period of comparative quiet followed, during which new seas were formed, and new life perhaps began to flourish in the lunar world, only to end in another cataclysm, which finally put a term to the existence of the moon as a life-supporting world.

Suppose we examine two more of Mr. Ritchey's illuminating photographs, and, first, the one showing the crater Theophilus and its surroundings. We have spoken of Theophilus before, citing the facts that it is sixty-four miles in diameter and eighteen thousand feet deep. It will be noticed that it has two brother giants — Cyrillus the nearer, and Catharina the more distant; but Theophilus is plainly the youngest of the trio. Centuries, and perhaps thousands of years, must have elapsed between the periods of their upheaval, for the two older craters are partly filled with débris, while it is manifest at a glance that when the southeastern wall of Theophilus was formed, it broke away and destroyed a part of the more ancient ring of Cyrillus. There is no more tremendous scene on the moon than this; viewed with a powerful telescope, it is absolutely appalling.

The next photograph shows, if possible, a still wilder region. It is the part of the moon lying between Tycho and the south pole. Tycho is seen in the lower left-hand part of the picture. To the right, at the edge of the illuminated portion of the moon, are the crater-rings, Longomontanus and Wilhelm I, the former being the larger. Between them are to be seen the ruins of two or three more ancient craters which, together with portions of the walls of Wilhelm I and Longomontanus, have been honeycombed with smaller craters. The vast crateriform depression above the center of the picture is Clavius, an unrivaled wonder of lunar scenery, a hundred and forty-two miles in its greatest length, while its whole immense floor has sunk two miles below the general surface of the moon outside the ring. The monstrous shadow-filled cavity above Clavius toward the right is Blancanus, whose aspect here gives a good idea of the appearance of these chasms when only their rims are in the sunlight. But observe the indescribable savagery of the entire scene. It looks as though the spirit of destruction had gone mad in this spot. The mighty craters have broken forth one after another, each rending its predecessor; and when their work was finished, a minor but yet tremendous outbreak occurred, and the face of the moon was gored and punctured with thousands of smaller craters. These relatively small craters (small, however, only in a lunar sense, for many of them would appear gigantic on the earth) recall once more the theory of meteoric impact. It does not seem impossible that some of them may have been formed by such an agency.

One would not wish for our planet such a fate as that which has overtaken the moon, but we cannot be absolutely sure that something of the kind may not be in store for it. We really know nothing of the ultimate causes of volcanic activity, and some have suggested that the internal energies of the earth may be accumulating instead of dying out, and may never yet have exhibited their utmost destructive power. Perhaps the best assurance that we can find that the earth will escape the catastrophe that has overtaken its satellite is to be found in the relatively great force of its gravitation. The moon has been the victim of its weakness; given equal forces, and the earth would be the better able to withstand them. It is significant, in connection with these considerations, that the little planet Mercury, which seems also to have parted with its air and water, shows to the telescope some indications that it is pitted with craters resembling those that have torn to pieces the face of the moon.

Upon the whole, after studying the dreadful lunar land-scapes, one cannot feel a very enthusiastic sympathy with those who are seeking indications of the continued existence of some kind of life on the moon; such a world is better without inhabitants. It has met its fate; let it go! Fortunately, it is not so near that it cannot hide its scars and appear beautiful — except when curiosity impels us to look with the penetrating eyes of the astronomer.

The Great Mars Problem

*L*et any thoughtful person who is acquainted with the general facts of astronomy look up at the heavens some night when they appear in their greatest splendor, and ask himself what is the strongest impression that they make upon his mind. He may not find it easy to frame an answer, but when he has succeeded it will probably be to the effect that the stars give him an impression of the universality of intelligence; they make him feel, as the sun and the moon cannot do, that his world is not alone; that all this was not made simply to form a gorgeous canopy over the tents of men. If he is of a devout turn of mind, he thinks, as he gazes into those fathomless deeps and among those bewildering hosts, of the infinite multitude of created beings that the Almighty has taken under his care. The narrow ideas of the old geocentric theology, which made the earth God's especial footstool, and man his only rational creature, fall away from him like a veil that had obscured his vision; they are impossible in the presence of what he sees above. Thus the natural tendency, in the light of modern progress, is to regard the universe as everywhere filled with life.

But science, which is responsible for this broadening of men's thoughts concerning the universality of life, itself proceeds to set limits. Of spiritual existences it pretends to know nothing, but as to physical beings, it declares that it can only entertain the supposition of their existence where it finds evidence of an environment suited to their needs, and such environment may not everywhere exist. Science, though repelled by the antiquated theological conception of the supreme isolation of man among created beings, regards with complacency the probability that there are regions in the universe where no organic life exists, stars which shine upon no inhabited worlds, and planets which nourish no animate creatures. The astronomical view of the universe is that it consists of matter in every stage of evolution: some nebulous and chaotic; some just condensing into stars (suns) of every magnitude and order; some shaped into finished solar bodies surrounded by dependent planets; some forming stars that perhaps have no planets, and will have none; some constituting suns that are already aging, and will soon lose their radiant energy and disappear; and some aggregated into masses that long ago became inert, cold, and rayless, and that can only be revivified by means about which we can form conjectures, but of which we actually know nothing.

As with the stars, so with the planets, which are the satellites of stars. All investigations unite to tell us that the planets are not all in the same state of development. As some are large and some small, so some are, in an evolutionary sense, young, and some old. As they depend upon the suns around which they revolve for their light, heat, and other forms of radiant energy, so their condition varies with their distance from those suns. Many may never arrive at a state suitable for the maintenance of life upon their surfaces; some which are not at present in such a state may attain it later; and the forms of life themselves may vary with the peculiar environment that different planets afford. Thus we see that we are not scientifically justified in affirming that life is ubiquitous, although we are thus justified in saying that it must be, in a general sense, universal. We might liken the universe to a garden known to contain every variety of plant. If on entering it we see no flowers, we examine the species before us and find that they are not of those which bloom at this particular season, or perhaps they are such as never bear flowers. Yet we feel no doubt that we shall find flowers somewhere in the garden, because there *are* species which bloom at this season, and the garden contains *all* varieties.

While it is tacitly assumed that there are planets revolving around other stars than the sun, it would be impossible for us to see them with any telescope yet invented, and no instrument now in the possession of astronomers could assure us of their existence; so the only planetary system of which we have visual knowledge is our own. Excluding the asteroids, which could not from any point of view be considered as habitable, we have in the solar system eight planets of various sizes and situated at various distances from the sun. Of these eight we know that one, the earth, is inhabited. The question, then, arises: Are there any of the others which are inhabited or habitable? Since it is our intention to discuss the habitability of only one of the seven to which the question applies, the rest may be dismissed in a few words. The smallest of them, and the nearest to the sun, is Mercury, which is regarded as uninhabitable because it has no perceptible supply of water and air, and because, owing to the extraordinary eccentricity of its orbit, it is subjected to excessive and very rapid alterations in the amount of solar heat and light poured upon its surface, such alterations being inconsistent with the supposition that it can support living beings. Even its average temperature is more than six and a half times that prevailing on the earth! Another circumstance which militates against its habitability is that, according to the results of the best telescopic studies, it always keeps the same face toward the sun, so that one half of the planet is perpetually exposed to the fierce solar rays, and the other half faces the unmitigated cold of open space. Venus, the next in distance from the sun, is almost the exact twin of the earth in size, and many arguments may be urged in favor of its habitability, although it is suspected of possessing the same peculiarity as Mercury, in always keeping the same side sunward. Unfortunately its atmosphere appears to be so dense that no permanent markings on its surface are certainly visible, and the question of its actual condition must, for the present, be left in abeyance. Mars, the first planet more distant from the sun than the earth, is the special subject of this chapter, and will be described and discussed a few lines further on. Jupiter, Saturn, Uranus, and Neptune, the four giant planets, all more distant than Mars, and each more distant than the other in the order named, are all regarded as uninhabitable because none of them appears to possess any degree of solidity. They may have solid or liquid nuclei, but exteriorly they seem to be mere balls of cloud. Of course, one can imagine what he pleases about the existence of creatures suited to the physical constitution of such planets as these, but they must be excluded from the category of habitable worlds in the ordinary sense of the term. We go back, then, to Mars.

It will be best to begin with a description of the planet. Mars is 4230 miles in diameter; its surface is not much more than one-quarter as extensive as that of the earth (.285). Its mean distance from the sun is 141,500,000 miles, 48,500,000 miles greater than that of the earth. Since radiant energy varies inversely as the square of distance, Mars receives less than half as much solar light and heat as the earth gets. Mars' year (period of revolution round the sun) is 687 days. Its mean density is 71 percent of the earth's, and the force of gravity on its surface is 38 percent of that on the surface of the earth; *i.e.,* a body weighing one hundred pounds on the earth would, if transported to Mars, weigh but thirty-eight pounds. The inclination of its equator to the plane of its orbit differs very little from that of the earth's equator, and its axial rotation occupies 24 hours 37 minutes. so that the length of day and night, and the extent of the seasonal changes on Mars, are almost precisely the same as on the earth. But owing to the greater length of its year, the seasons of Mars, while occurring in the same order, are almost twice as long as ours. The surface of the planet is manifestly solid, like that of our globe, and the telescope reveals many permanent markings on it, recalling the appearance of a globe on which geographical features have been represented in reddish and dusky tints. Around the poles are plainly to be seen rounded white areas, which vary in extent with the Martian seasons, nearly vanishing in summer and extending widely in winter. The most recent spectroscopic determinations indicate that Mars has an atmosphere perhaps as dense as that to be found on our loftiest mountain peaks, and there is a perceptible amount of watery vapor in this atmosphere. The surface of the planet appears to be remarkably level, and it has no mountain ranges. No evidences of volcanic action have been discovered on Mars. The dusky and reddish areas were regarded by the early observers as respectively seas and lands, but at present it is not believed that there are any bodies of water on the planet. There has never been much doubt expressed that the white areas about the poles represent snow.

It will be seen from this brief description that many remarkable resemblances exist between Mars and the earth, and there is nothing wonderful in the fact that the question of the habitability of the former has become one of extreme and wide-spread interest, giving rise to the most diverse views, to many extraordinary speculations, and sometimes to regrettably heated controversy. The first champion of the habitability of Mars was Sir William Herschel, although even before his time the idea had been suggested. He was convinced by the revelations of his telescopes, continually increasing in power, that Mars was more like the earth than any other planet. He could not resist the testimony of the polar snows, whose suggestive conduct was in such striking accord with what occurs upon the earth. Gradually, as telescopes improved and observers increased in number, the principal features of the planet were disclosed and charted, and "areography," as the geography of Mars was called, took its place among the recognized branches of astronomical study. But it was not before 1877 that a fundamentally new discovery in areography gave a truly sensational turn to speculation about life on "the red planet." In that year Mars made one of its nearest approaches to the earth, and was so situated in its orbit that it could be observed to great advantage from the northern hemisphere of the earth. The celebrated Italian astronomer, Schiaparelli, took advantage of this opportunity to make a trigonometrical survey of the surface of Mars — as coolly and confidently as if he were not taking his sights across a thirty-five-million-mile gulf of empty space — and in the course of this survey he was astonished to perceive that the reddish areas, then called continents, were crossed in many directions by narrow, dusky lines, to which he gave the suggestive name of "canals." Thus a kind of firebrand was cast into the field of astronomical speculation, which has ever since produced disputes that have sometimes approached the violence of political faction. At first the accuracy of Schiaparelli's observations was contested; it required a powerful telescope, and the most excellent "seeing," to render the enigmatical lines visible at all, and many searchers were unable to detect them.

But Schiaparelli continued his studies in the serene sky of Italy, and produced charts of the gridironed face of Mars containing so much astonishing detail that one had either to reject them *in toto* or to confess that Schiaparelli was right. As subsequent favorable oppositions of Mars occurred, other observers began to see the "canals" and to confirm the substantial accuracy of the Italian astronomer's work, and finally few were found who would venture to affirm that the "canals" did not exist, whatever their meaning might be.

When Schiaparelli began his observations it was generally believed, as we have said, that the dusky areas on Mars were seas, and since Schiaparelli thought that the "canals" invariably began and ended at the shores of the "seas," the appropriateness of the title given to the lines seemed apparent. Their artificial character was immediately assumed by many, because they were too straight and too suggestively geometrical in their arrangement to permit the conclusion that they were natural watercourses. A most surprising circumstance noted by Schiaparelli was that the "canals" made their appearance *after* the melting of the polar snow in the corresponding hemisphere had begun, and that they grew darker, longer, and more numerous in proportion as the polar liquidation proceeded; another very puzzling observation was that many of them became double as the season advanced; close beside an already existing "canal," and in perfect parallelism with it, another would gradually make its appearance. That these phenomena actually existed and were not illusions was proved by later observations, and today they are seen whenever Mars is favorably situated for observation.

In the closing decade of the nineteenth century, Mr. Percival Lowell took up the work where Schiaparelli had virtually dropped it, and soon added a great number of "canals" to those previously known, so that in his charts the surface of the wonderful little planet appears covered as with a spider's web, the dusky lines crisscrossing in every direction, with conspicuous knots wherever a number of them come together. Mr. Lowell has demonstrated that the areas originally called seas, and thus named on the earlier charts, are not bodies of water, whatever else they may be. He has also found that the mysterious lines do not, as Schiaparelli supposed, begin and end at the edges of the dusky regions, but often continue on across them, reaching in some cases far up into the polar regions. But Schiaparelli was right in his observation that the appearance of the "canals" is synchronous with the gradual disappearance of the polar snows, and this fact has become the basis of the most extraordinary theory that the subject of life in other worlds has ever given birth to.

Now, the effect of such discoveries, as we have related, depends upon the type of mind to whose attention they are called. Many are content to accept them as strange and inexplicable at present, and to wait for further light upon them; others insist upon an immediate inquiry concerning their probable nature and meaning. Such an inquiry can only be based upon inference proceeding from analogy. Mars, say Mr. Lowell and those who are of his opinion, is manifestly a solidly incrusted planet like the earth; it has an atmosphere, though one of great rarity; it has water vapor, as the snows in themselves prove; it has the alternation of day and night, and a succession of seasons closely resembling those of the earth; its surface is suggestively divided into regions of contrasting colors and appearance, and upon that surface we see an immense number of lines geometrically arranged, with a system of symmetrical intersections where the lines expand into circular and oval areas — and all connected with the annual melting of the polar snows in a way which irresistibly suggests the interference of intelligence directed to a definite end. Why, with so many concurrent circumstances to support the hypothesis, should we not regard Mars as an inhabited globe?

But the differences between Mars and the earth are in many ways as striking as their resemblances. Mars is relatively small; it gets less than half as much light and heat as we receive; its atmosphere is so rare that it would be distressing to us, even if we could survive in it at all; it has no lakes, rivers, or seas; its surface is an endless prairie. and its "canals" are phenomena utterly unlike anything on the earth. Yet it is precisely upon these divergences between the earth and Mars, this repudiation of terrestrial standards, that the theory of "life on Mars," for which Mr. Lowell is mainly responsible, is based. Because Mars is smaller than the earth, we are told it must necessarily be more advanced in planetary evolution, the underlying cause of which is the gradual cooling and contraction of the planet's mass. Mars has parted with its internal heat more rapidly than the earth; consequently its waters and its atmosphere have been mostly withdrawn by chemical combinations, but enough of both yet remain to render life still possible on its surface. As the globe of Mars is evolutionally older than that of the earth, so its forms of organic life may be proportionally further advanced, and its inhabitants may have attained a degree of cultivated intelligence much superior to what at present exists upon the earth. Understanding the nature and the causes of the desiccation of their planet, and possessing engineering science and capabilities far in advance of ours, they may be conceived to have grappled with the stupendous problem of keeping their world in a habitable condition as long as possible. Supposing them to have become accustomed to live in their rarefied atmosphere (a thing not inconceivable, since men can live for a time at least in air hardly less rare), the most pressing problem for them is that of a water-supply, without which plant life cannot exist, while animal life in turn depends for its existence upon vegetation. The only direction in which they can seek water is that of the polar regions, where it is alternately condensed into snow and released in the liquid form by the effect of the seasonal changes. It is, then, to the annual melting of the polar snow-fields that the Martian engineers are supposed to have recourse in supplying the needs of their planet, and thus providing the means of prolonging their own existence. It is imagined that they have for this purpose constructed a stupendous system of irrigation extending over the temperate and equatorial regions of the planet. The "canals" represent the lines of irrigation, but the narrow streaks that we see are not the canals themselves, but the irrigated bands covered by them. Their dark hue, and their gradual appearance after the polar melting has begun, are due to the growth of vegetation stimulated by the water. The rounded areas visible where several "canals" meet and cross are called by Mr. Lowell "oases." These are supposed to be the principal centers of population and industry. It must be confessed that some of them, with their complicated systems of radiating lines, appear to answer very well to such a theory. No attempt to explain them by analogy with natural phenomena on the earth has proved successful.

But a great difficulty yet remains: How to explain the seemingly miraculous powers of the supposed engineers? Here recourse is had once more to the relative smallness of the planet. We have remarked that the force of gravity on Mars is only thirty-eight percent of that on the earth. A steam-shovel driven by a certain horse-power would be nearly three times as effective there as here. A man of our stature on Mars would find his effective strength increased in the same proportion. But just because of the slight force of gravity there, a Martian might attain to the traditional stature of Goliath without finding his own weight an encumbrance to his activity, while at the same time his huge muscles would come into unimpeded play, enabling him single-handed to perform labors that would be impossible to a whole gang of terrestrial workmen. The effective powers of huge machines would be increased in the same way; and to all this must be added the fact that the mean density of the materials of which Mars is composed is much less than that of the constituents of the earth. Combining all these considerations, it becomes much less difficult to conceive that public works might be successfully undertaken on Mars which would be hopelessly beyond the limits of human accomplishment.

Certain other difficulties have also to be met; as, for instance, the relative coldness of the climate of Mars. At its distance it gets considerably less than half as much light and heat as we receive. In addition to this, the rarity of its atmosphere would naturally be expected to decrease the effective temperature at the planet's surface, since an atmosphere acts somewhat like the glass cover of a hothouse in retaining the solar heat which has penetrated it. It has been calculated that, unless there are mitigating circumstances of which we know nothing, the average temperature at the surface of Mars must be far below the freezing-point of water. To this it is replied that the possible mitigating circumstances spoken of evidently exist in fact, because we can *see* that the watery vapor condenses into snow around the poles in winter, but melts again when summer comes. The mitigating agent may be supposed to exist in the atmosphere where the presence of certain gases would completely alter the temperature gradients.

It might also be objected that it is inconceivable that the Martian engineers, however great may be their physical powers, and however gigantic the mechanical energies under their control, could force water in large quantities from the poles to the equator. This is an achievement that measures up to the cosmical standard. It is admitted by the champions of the theory that the difficulty is a formidable one; but they call attention to the singular fact that on Mars there can be found no chains of mountains, and it is even doubtful if ranges of hills exist there. The entire surface of the planet appears to be almost "as smooth as a billiard ball," and even the broad regions which were once supposed to be seas apparently lie at practically the same level as the other parts, since the "canals" in many cases run uninterruptedly across them. Lowell's idea is that these somber areas may be expanses of vegetation covering ground of a more or less marshy character, for while the largest of them appear to be permanent, there are some which vary coincidently with the variations of the canals.

As to the kind of machinery employed to force the water from the poles, it has been conjectured that it may have taken the form of a gigantic system of pumps and conduits; and since the Martians are assumed to be so far in advance of us in their mastery of scientific principles, the hypothesis will at least not be harmed by supposing that they have learned to harness forces of nature whose very existence in a manageable form is yet unrecognized on the earth. If we wish to let the imagination loose, we may conjecture that they have conquered the secret of those intra-atomic forces whose resistless energy is beginning to become evident to us, but the possibility of whose utilization remains a dream, the fulfillment of which nobody dares to predict.

Such, in very brief form, is the celebrated theory of Mars as an inhabited world. It certainly captivates the imagination, and if we believe it to represent the facts, we cannot but watch with the deepest sympathy this gallant struggle of an intellectual race to preserve its planet from the effects of advancing age and death. We may, indeed, wonder whether our own humanity, confronted by such a calamity, could be counted on to meet the emergency with equal stoutness of heart and inexhaustibleness of resource. Up to the present time we certainly have shown no capacity to confront Nature toe to toe, and to seize her by the shoulders and turn her round when she refuses to go our way. If we could get into wireless telephonic communication with the Martians we might learn from their own lips the secret of their more than "Roman recovery."

The Riddle of the Asteroids

*B*etween the orbits of Mars and Jupiter revolves the most remarkable system of little bodies with which we are acquainted — the Asteroids, or Minor Planets. Some six hundred are now known, and they may actually number thousands. They form virtually a ring about the sun. The most striking general fact about them is that they occupy the place in the sky which should be occupied, according to Bode's Law, by a single large planet. This fact, as we shall see, has led to the invention of one of the most extraordinary theories in astronomy — *viz.,* that of the explosion of a world!

Bode's Law, so-called, is only an empiric formula, but until the discovery of Neptune it accorded so well with the distances of the planets that astronomers were disposed to look upon it as really representing some underlying principle of planetary distribution. They were puzzled by the absence of a planet in the space between Mars and Jupiter, where the "law" demanded that there should be one, and an association of astronomers was formed to search for it. There was a decided sensation when, in 1801, Piazzi, of Palermo, announced that he had found a little planet which apparently occupied the place in the system which belonged to the missing body. He named it Ceres, and it was the first of the Asteroids. The next year Olbers, of Bremen, while looking for Ceres with his telescope, stumbled upon another small planet which he named Pallas. Immediately he was inspired with the idea that these two planets were fragments of a larger one which had formerly occupied the vacant place in the planetary ranks, and he predicted that others would be found by searching in the neighborhood of the intersection of the orbits of the two already discovered. This bold prediction was brilliantly fulfilled by the finding of two more — Juno in 1804, and Vesta in 1807. Olbers would seem to have been led to the invention of his hypothesis of a planetary explosion by the faith which astronomers at that time had in Bode's Law. They appear to have thought that several planets revolving in the gap where the "law" called for but one could only be accounted for upon the theory that the original *one* had been broken up to form the several. Gravitation demanded that the remnants of a planet blown to pieces, no matter how their orbits might otherwise differ, should all return at stated periods to the point where the explosion had occurred; hence Olbers' prediction that any asteroids that might subsequently be discovered would be found to have a common point of orbital intersection. And curiously enough all of the first asteroids found practically answered to this requirement. Olbers' theory seemed to be established.

After the first four, no more asteroids were found until 1845, when one was discovered; then, in 1847, three more were added to the list; and after that searchers began to pick them up with such rapidity that by the close of the century hundreds were known, and it had become almost impossible to keep track of them. The first four are by far the largest members of the group, but their actual sizes remained unknown until less than twenty years ago. It was long supposed that Vesta was the largest, because it shines more brightly than any of the others; but finally, in 1895, Barnard, with the Lick telescope, definitely measured their diameters, and proved to everybody's surprise that Ceres is really the chief, and Vesta only the third in rank. His measures are as follows: Ceres, 477 miles; Pallas, 304 miles; Vesta, 239 miles; and Juno, 120 miles. They differ greatly in the reflective power of their surfaces, a fact of much significance in connection with the question of their origin. Vesta is, surface for surface, rather more than three times as brilliant as Ceres, whence the original mistake about its magnitude.

Nowadays new asteroids are found frequently by photography, but physically they are most insignificant bodies, their average diameter probably not exceeding twenty miles, and some are believed not to exceed ten. On a planet only ten miles in diameter, assuming the same mean density as the earth's, which is undoubtedly too much, the force of gravity would be so slight that an average man would not weigh more than three ounces, and could jump off into space whenever he liked.

Although the asteroids all revolve around the sun in the same direction as that pursued by the major planets, their orbits are inclined at a great variety of angles to the general plane of the planetary system, and some of them are very eccentric — almost as much so as the orbits of many of the periodic comets. It has even been conjectured that the two tiny moons of Mars and the four smaller satellites of Jupiter may be asteroids gone astray and captured by those planets. Two of the asteroids are exceedingly remarkable for the shapes and positions of their orbits; these are Eros, discovered in 1898, and T.G., 1906, found eight years later. The latter has a mean distance from the sun slightly greater than that of Jupiter, while the mean distance of Eros is less than that of Mars. The orbit of Eros is so eccentric that at times it approaches within 15,000,000 miles of the earth, nearer than any other regular member of the solar system except the moon, thus affording an unrivaled means of measuring the solar parallax. But for our present purpose the chief interest of Eros lies in its extraordinary changes of light.

These changes, although irregular, have been observed and photographed many times, and there seems to be no doubt of their reality. Their significance consists in their possible connection with the form of the little planet, whose diameter is generally estimated at not more than twenty miles. Von Oppolzer found, in 1901, that Eros lost three-fourths of its brilliancy once in every two hours and thirty-eight minutes. Other observers have found slightly different periods of variability, but none as long as three hours. The most interesting interpretation that has been offered of this phenomenon is that it is due to a great irregularity of figure, recalling at once Olbers' hypothesis. According to some, Eros may be double, the two bodies composing it revolving around each other at very close quarters; but a more striking, and it may be said probable, suggestion is that Eros has a form not unlike that of a dumb-bell, or hour-glass, turning rapidly end over end so that the area of illuminated surface presented to our eyes continually changes, reaching at certain times a minimum when the amount of light that it reflects toward the earth is reduced to a quarter of its maximum value. Various other bizarre shapes have been ascribed to Eros, such, for instance, as that of a flat stone revolving about one of its longer axes, so that sometimes we see its face and sometimes its edge.

All of these explanations proceed upon the assumption that Eros cannot have a simple globular figure like that of a typical planet, a figure which is prescribed by the law of gravitation, but that its shape is what may be called accidental; in a word, it is a *fragment,* for it seems impossible to believe that a body formed in interplanetary space, either through nebular condensation or through the aggregation of particles drawn together by their mutual attractions, should not be practically spherical in shape. Nor is Eros the only asteroid that gives evidence by variations of brilliancy that there is something abnormal in its constitution; several others present the same phenomenon in varying degrees. Even Vesta was regarded by Olbers as sufficiently variable in its light to warrant the conclusion that it was an angular mass instead of a globe. Some of the smaller ones show very notable variations, and all in short periods, of three or four hours, suggesting that in turning about one of their axes they present a surface of variable extent toward the sun and the earth.

The theory which some have preferred — that the variability of light is due to the differences of reflective power on different parts of the surface — would, if accepted, be hardly less suggestive of the origin of these little bodies by the breaking up of a larger one, because the most natural explanation of such differences would seem to be that they arose from variations in the roughness or smoothness of the reflecting surface, which would be characteristic of fragmentary bodies. In the case of a large planet alternating expanses of land and water, or of vegetation and desert, would produce a notable variation in the amount of reflection, but on bodies of the size of the asteroids neither water nor vegetation could exist, and an atmosphere would be equally impossible.

One of the strongest objections to Olbers' hypothesis is that only a few of the first asteroids discovered travel in orbits which measurably satisfy the requirement that they should all intersect at the point where the explosion occurred. To this it was at first replied that the perturbations of the asteroidal orbits, by the attractions of the major planets, would soon displace them in such a manner that they would cease to intersect. One of the first investigations undertaken by the late Prof. Simon Newcomb was directed to the solution of this question, and he arrived at the conclusion that the planetary perturbations could not explain the actual situation of the asteroidal orbits. But afterward it was pointed out that the difficulty could be avoided by supposing that not one but a series of explosions had produced the asteroids as they now are. After the primary disruption the fragments themselves, according to this suggestion, may have exploded, and then the resulting orbits would be as "tangled" as the heart could wish. This has so far rehabilitated the explosion theory that it has never been entirely abandoned, and the evidence which we have just cited of the probably abnormal shapes of Eros and other asteroids has lately given it renewed life. It is a subject that needs a thorough rediscussion.

We must not fail to mention, however, that there is a rival hypothesis which commends itself to many astronomers — *viz.*, that the asteroids were formed out of a relatively scant ring of matter, situated between Mars and Jupiter and resembling in composition the immensely more massive rings from which, according to Laplace's hypothesis, the planets were born. It is held by the supporters of this theory that the attraction of the giant Jupiter was sufficient to prevent the small, nebulous ring that gave birth to the asteroids from condensing like the others into a single planet.

But if we accept the explosion theory, with its corollary that minor explosions followed the principal one, we have still an unanswered question before us: What caused the explosions? The idea of *a world blowing up* is too Titanic to be shocking; it rather amuses the imagination than seriously impresses it; in a word, it seems essentially chimerical. We can by no appeal to experience form a mental picture of such an occurrence. Even the moon did not blow up when it was wrecked by volcanoes. The explosive nebulæ and new stars are far away in space, and suggest no connection with such a catastrophe as the bursting of a planet into hundreds of pieces. We cannot conceive of a great globe thousands of miles in diameter resembling a pellet of gunpowder only awaiting the touch of a match to cause its sudden disruption. Somehow the thought of human agency obtrudes itself in connection with the word "explosion," and we smile at the idea that giant powder or nitroglycerin could blow up a planet. Yet it would only need *enough* of them to do it.

After all, we may deceive ourselves in thinking, as we are apt to do, that explosive energies lock themselves up only in small masses of matter. There are many causes producing explosions in nature, every volcanic eruption manifests the activity of some of them. Think of the giant power of confined steam; if enough steam could be suddenly generated in the center of the earth by a downpour of all the waters of the oceans, what might not the consequences be for our globe? In a smaller globe, and it has never been estimated that the original asteroid was even as large as the moon, such a catastrophe would, perhaps, be more easily conceivable; but since we are compelled in this case to assume that there was a series of successive explosions, steam would hardly answer the purpose; it would be more reasonable to suppose that the cause of the explosion was some kind of chemical reaction, or something affecting the atoms composing the exploding body. Here Dr. Gustav Le Bon comes to our aid with a most startling suggestion, based on his theory of the dissipation of intra-atomic energy. It will be best to quote him at some length from his book on *The Evolution of Forces.*

"It does not seem at first sight," says Doctor Le Bon,

very comprehensible that worlds which appear more and more stable as they cool could become so unstable as to afterward dissociate entirely. To explain this phenomenon, we will inquire whether astronomical observations do not allow us to witness this dissociation.

We know that the stability of a body in motion, such as a top or a bicycle, ceases to be possible when its velocity of rotation descends below a certain limit. Once this limit is reached it loses its stability and falls to the ground. Prof. J.J. Thomson even interprets radio-activity in this manner, and points out that when the speed of the elements composing the atoms descends below a certain limit they become unstable and tend to lose their equilibria. There would result from this a commencement of dissociation, with diminution of their potential energy and a corresponding increase of their kinetic energy sufficient to launch into space the products of intra-atomic disintegration.

It must not be forgotten that the atom being an enormous reservoir of energy is by this very fact comparable with explosive bodies. These last remain inert so long as their internal equilibria are undisturbed. So soon as some cause or other modifies these, they explode and smash everything around them after being themselves broken to pieces.

Atoms, therefore, which grow old in consequence of the diminution of a part of their intra-atomic energy gradually lose their stability. A moment, then, arrives when this stability is so weak that the matter disappears by a sort of explosion more or less rapid. The bodies of the radium group offer an image of this phenomenon – a rather faint image, however, because the atoms of this body have only reached a period of instability when the dissociation is rather slow. It probably precedes another and more rapid period of dissociation capable of producing their final explosion. Bodies such as radium, thorium, etc., represent, no doubt, a state of old age at which all bodies must some day arrive, and which they already begin to manifest in our universe, since all matter is slightly radioactive. It would suffice for the dissociation to be fairly general and fairly rapid for an explosion to occur in a world where it was manifested.

These theoretical considerations find a solid support in the sudden appearances and disappearances of stars. The explosions of a world which produce them reveal to us, perhaps, how the universes perish when they become old.

As astronomical observations show the relative frequency of these rapid destructions, we may ask ourselves whether the end of a universe by a sudden explosion after a long period of old age does not represent its most general ending.

Here, perhaps, it will be well to stop, since, entrancing as the subject may be, we know very little about it, and Doctor Le Bon's theory affords a limitless field for the reader's imagination.